Mesenchymal Stem Cells and Skeletal Regeneration

Mesenchymal Stem Cells and Skeletal Regeneration

Elena A. Jones, PhD
Associate Professor
Academic Unit of Musculoskeletal Disease
Leeds Institute of Molecular Medicine
St. James's University Hospital
Leeds, UK

Xuebin Yang, Med., MSc (Hand, Orth), PhD (Stem cell, TE)
Head of Tissue Engineering Research
Biomaterials and Tissue Engineering Group
Leeds Dental Institute
University of Leeds
Leeds, UK

Peter Giannoudis, MD FRCS
Professor — Section Head
Academic Department of Trauma & Orthopaedic Surgery
School of Medicine
University of Leeds
Leeds, UK

Dennis McGonagle, PhD, FRCPI, NIHR
Leeds Musculoskeletal Biomedical Research Unit
Leeds Institute of Molecular Medicine
University of Leeds
Leeds, UK

ELSEVIER

AMSTERDAM • BOSTON • HEIDELBERG • LONDON
NEW YORK • OXFORD • PARIS • SAN DIEGO
SAN FRANCISCO • SINGAPORE • SYDNEY • TOKYO
Academic Press is an imprint of Elsevier

Academic Press is an imprint of Elsevier
The Boulevard, Langford Lane, Kidlington, Oxford, OX5 1GB, UK
225 Wyman Street, Waltham, MA 02451, USA

First published 2013

British Library Cataloguing in Publication Data
A catalogue record for this book is available from the British Library

Library of Congress Cataloging-in-Publication Data
A catalog record for this book is available from the Library of Congress

ISBN: 978-0-12-407915-1

For information on all Academic Press publications
visit our website at store.elsevier.com

This book has been manufactured using Print On Demand technology. Each copy is produced to order and is limited to black ink. The online version of this book will show color figures where appropriate.

Working together to grow
libraries in developing countries

www.elsevier.com | www.bookaid.org | www.sabre.org

ELSEVIER   BOOK AID
            International   Sabre Foundation

Transferred to Digital Printing in 2013

CONTENTS

CHAPTER 1

Current Strategies for Skeletal Regeneration in the Early Twenty-First Century

1.1 THE ECONOMIC AND SOCIAL BURDEN OF DISEASES AFFECTING BONE AND CARTILAGE

There is a growing number of clinical conditions, in which the normal process of skeletal regeneration is impaired. In health, the maintenance of stable bone mass is the result of a carefully controlled balance between the activities of bone-forming osteoblasts and bone-resorbing osteoclasts [1, 2]. In the clinical setting, the most common form of intrinsic bone regeneration is fracture healing; however, approximately 10% of fractures fail to heal and require additional interventions [3]. In orthopedic surgery, there are other conditions that require bone regeneration in high quantity, for example, bone reconstruction after tumor resection or after large loss of bone due to infection, trauma, or skeletal abnormality [4]. In dental and maxillofacial applications, the bone often needs strengthening prior to dental implant surgery and in some cases, large fragments of bone require "rebuilding" following injuries to the head. In some conditions, such as avascular necrosis, the innate regenerative process is compromised, leading to disability, with current surgical treatments failing to provide long-term improvements [4, 5].

Furthermore, there exist systemic bone abnormalities such as osteoporosis (OP), which is one of the commonest diseases among older females. It is characterized by disequilibrium of bone formation and resorption, leading to weakening of bone, which in turn contributes to the increased risk of fractures [6]. With mortality rates of 30% at 1 year post injury, OP fragility fractures pose a great challenge to both social and insurance-based healthcare economies, with the annual UK costs alone estimated to be in a region of £2 billion [7]. Another very common musculoskeletal age-related disorder is osteoarthritis (OA), the disease affecting both bone and cartilage, which similarly causes considerable morbidity and mortality [8, 9]. With a prevalence of hip OA reported to be reaching 8% [10], indirect costs, due to OA-related workday absences, lead to significant loss in productivity in the

Jones: Mesenchymal Stem Cells and Skeletal Regeneration.

Western economies [11]. Currently, OA is treated only symphomatically, with the ultimate solution to alleviate pain being joint replacement surgery. Equally, despite recent advances in the treatment of OP using drugs targeting the osteoclastic lineage [4, 6], these drugs can lead to a low bone turnover state and reduced osteoblastic activity [12]; therefore, the disease remains a major health burden to the European economies.

1.2 CURRENT CLINICAL APPROACHES FOR BONE REGENERATION: THE DIAMOND CONCEPT

The current management of complex clinical situations in which the normal process of skeletal regeneration is impaired involves a number of treatment methods that alone, or in combination, lead to the enhancement of innate healing processes [13]. In relation to repairing bone, four equally-important factors have been proposed to be necessary and act in concert: adequate mechanical stability, osteoinductive agents such as growth factors, osteoconductive scaffolds, and osteogenic cells (the diamond concept) [14]. Standard approaches widely used in clinic, such as distraction osteogenesis/bone transport (Figure 1.1), primarily act via the biomechanical stimulation route [15]. Bone voids in large segmental bone defects can be also reconstructed with the use of bone autografts, most commonly from anterior or posterior iliac crests of the pelvis [4] (Figure 1.2). The fibula is another donor bone frequently used for the reconstruction of large defects in long bones; this is due to its size, configuration, and the ability to promote early remodeling [16] (Figure 1.3A). More recently, autograft material has been also harvested from the intramedullary canal of long bones using a new reaming system, called reaming−irrigator−aspirator [17]. To its advantage, large volumes of graft material can be collected; however, safety concerns and complications still exist pertaining to discomfort to the patient and high cost of the procedure [4]. Despite their known disadvantages, including pain and discomfort to the patient, autografts remain the "gold standard" in bone reconstruction as they encompass all four essential components of the diamond concept [14].

Allografts, including large femoral allografts (Figure 1.3B), can be used in limb salvage procedures after resection of aggressive bone tumor; revitalization and durability of the graft remain the most

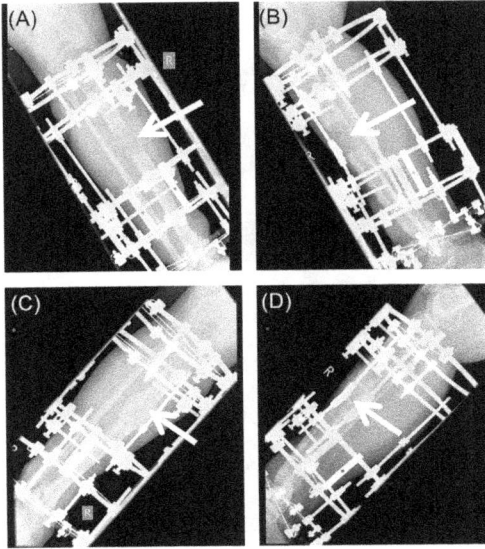

Figure 1.1 Technique of distraction osteogenesis illustrating the treatment of a 5 cm tibial bone defect using the Ilizarov fine wire fixator following proximal corticotomy. (A, B) Anterior–posterior and lateral radiographs at 3 months follow-up; (C, D) anterior–posterior and lateral radiographs at 6 months follow-up. White arrows show the regenerative area (A, B) and the maturation of the regenerative bone (C, D).

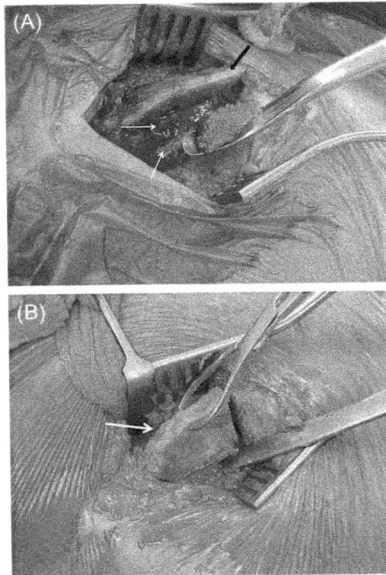

Figure 1.2 Harvesting of autologous cancellous (A) and tricortical (B, white arrow) bone grafts from anterior iliac crest. White arrows on the top image show cortical window made to get access to the inner cancellous graft.

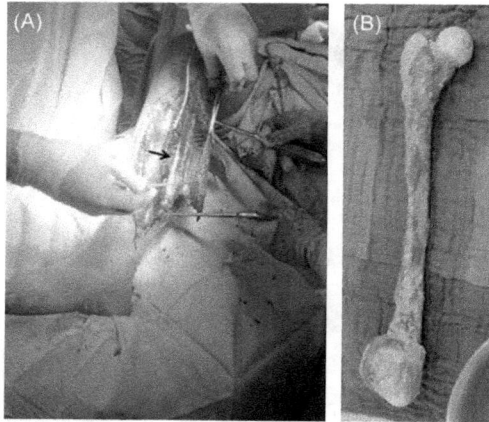

Figure 1.3 Harvesting of vascularized fibular graft from left lower extremity (A) and fresh femoral allograft (B).

significant concerns with this type of grafting [18]. Allogeneic bone is also available as demineralized bone matrix (DBM) and cancellous chips [4]. In the majority of allograft preparations, the cellular component is removed by irradiation or freeze-drying processes, in order to avoid immunogenicity [19]. In rare instances, however, allogeneic grafts may contain nonimmune cells from the donor; interestingly, such grafts have shown good safety in some clinical investigations [20, 21].

Bone-graft substitutes, commonly termed bone scaffolds, possess strong osteoconductive properties, that is, they are able to facilitate cell infiltration, maturation toward osteoblasts, and eventually extra-cellular matrix (ECM) deposition on the scaffold surface [22]. Natural scaffolds are commonly made from devitalized bovine bone, whereas the so-called synthetic scaffolds are based on biomaterials such as hydroxyapatite, beta-tricalcium phosphate, or glass ceramics [22−24]. The ideal scaffold should be biocompatible and nontoxic, offer bio-mechanical properties of the replacement bone, and should tolerate sterilization and reshaping to the required dimensions [22]. New-generation scaffolds based on synthetic polymers offer an additional advantage of being "biodegradable"; their degradation rate can be modified to be near to that of normal bone by copolymerization and changes in hydrophobicity [22, 24, 25].

The third component of a diamond concept pertains to osteogenic growth factors. The two growth factors most used in repairing bone are bone morphogenetic proteins (BMPs) 2 and 7. The first is manu-factured by Medtronics as a recombinant BMP-2 embedded in an absorbable collagen sponge (INFUSE). These two BMPs have been comprehensively evaluated in many clinical studies of nonunion frac-ture, open fracture, and spinal fusion [26]. Some surgeons however remain unconvinced that these BMPs enhance fracture healing to a sig-nificant degree [27]. Large clinical trials with BMP-2 and -7, including all appropriate control groups, are still missing and health economics analysis suggests that BMP treatment may be favorable economically only when used in patients with the most severe fractures [28]. Furthermore, some concerns have been recently raised regarding BMP-2 safety in spinal surgery [29].

The osteogenic cells are the most important component of the dia-mond concept of bone healing. A variety of terms are currently used to refer to these regenerative cells including skeletal stem cells [30, 31], osteoprogenitors [32, 33] or self-renewing osteoprogenitors [34], mar-row stromal cells [35−37], marrow stromal stem cells [38], mesenchy-mal progenitors [39, 40], and, finally, mesenchymal stem cells (MSCs) [41, 42]. The terminology has evolved in parallel with the knowledge on the biology of these cells, the issue which is discussed in the next chapter.

CHAPTER 2

Mesenchymal Stem Cells: Discovery in Bone Marrow and Beyond

2.1 DISCOVERY

The history of the biology of Mesenchymal Stem Cells (MSCs) owes it conception and birth to the earlier discovery of its more illustrious bone marrow (BM) resident sibling—the hematopoietic stem cell (HSC), whose existence was first proposed by Maximov in 1909. By 1960, there was a great interest in applying new emergent knowledge on HSCs toward BM transplantation strategies. *In vitro* cell culture and subsequently *in vivo* animal model assays showed the potential to transplant freshly isolated HSCs into primary and secondary recipients [43]. Based on these assays, new definitions of stem cells emerged, the main defining principle being an ability of a stem cell to self-renew, as illustrated by the HSCs to fully reconstitute hematopoiesis in secondary recipients. This new knowledge paved the way for the widespread adoption of BM transplantation in lymphoproliferative disease where aggressive antitumor ablation with irradiation and chemotherapy, followed by transplantation with healthy HSCs from tissue-compatible donors, could rescue the marrow [44].

In the Soviet Union, similar work was conducted at the Gamaleya's Institute in Moscow where scientists in Dr Alexander Friedenstain's group studied cells of BM microenvironment and their resistance to severe regimes of irradiation, a procedure used to "ablate" patient's marrow prior to transplantation [45]. In a course of their studies, Friedenstein et al. [46] noted that BM seeded in glass flasks and maintained in fairly basic culture media produced an intriguing adherent population of non-hematopoietic cells that formed colonies and were transplantable. Furthermore, this occurred at a single-colony level with transplanted colonies being capable of self-renewal and also forming mature nonhematopoietic tissues in recipient animals, such as bone, cartilage, and fibrous tissue thus indicating their potentially true stem cell nature [47].

Arnold Caplan and Darwin Prockop [35, 48] were among the first to recognize the importance of Friedentein's discoveries in the West. Additionally, Caplan proposed a concept of "mesengenesis," similar to a concept of hematopoiesis in the HSC field and was the first to coin the term "mesenchymal" stem cell (MSC) [48]. Below this ancestral stem cell, he placed a hierarchy of more mature "progenitor" cells, differentiation potentials of which were restricted to a narrower range of tissues, for example, osteo- and chondro-progenitors that gave rise to bone or cartilage tissues, respectively [49]. The first marker of a clonogenic marrow stromal cell, Stro-1, was described as early as 1991 [50]. Paulo Bianco and Pamela Robey [51] later proposed putative pericyte/reticular cell topography of MSCs in the BM.

The recognition that the high proliferative potential of culture-expanded MSCs offered an enormous potential for therapeutics in the fields of skeletal regeneration was first exploited by Osiris Therapeutics, a Baltimore-based biotechnology company, which brought MSCs to the forefront of scientific and public attention in a Science article entitled "Multilineage potential of adult human mesenchymal stem cells" [41]. This work synthesized the available knowledge from seemingly unrelated fields of bone, cartilage, and fat differentiation to develop novel *in vitro* functional assays for MSCs, which still form the basis for MSC characterization. Furthermore, the Science article proposed a set of markers to characterize MSCs retrospectively (i.e., following culture) and also provided first evidence of MSC heterogeneity at the single-cell level [41].

Finally, work emanating from the Osiris group and others also showed that culture-expanded MSCs appeared to be immunomodulatory and were capable of suppressing an array of inflammatory reactions [52]. Proof-of-concept studies of MSC immunomodulatory properties have been later performed using human cells [53]; shortly it proved to be instrumental in explaining some mechanisms of action by transplanted MSCs in many seemingly unrelated disease applications.

2.2 "EXPLOSION" OF INTEREST IN MSCs

It is hard to narrate more recent development of the MSC field without appearing "biased" as the field has literally "exploded," overtaking in terms of yearly publication volume its "parental" HSC field. Significant efforts were put toward the study of MSCs in different animal species to model MSC behavior in disease models and to establish a potential utility of MSC transplantation to treat these diseases. Such was the extent of MSC "plasticity" *in vitro*, that MSC-based experimental therapies have entered the fields of liver and neuronal tissue regeneration [54, 55], the tissues ontogenetically distant from the embryonic "mesenchyme." A snapshot of clinical trials involving MSC transplantation for various disease indications is given in Table 2.1.

In 2001, De Bari et al. [56] demonstrated the presence of MSCs in the synovium, a connective tissue lining the joint from inside and

Table 2.1. Recently Completed and Published Clinical Trials Utilizing MSC from Different Tissue Sources

Disease Category	Example of a Trial/Trial Identifier	Trial Type	Type of MSCs Used	Sponsor
Heart and Blood Diseases	POSEIDON[a] NCT01087996 [318]	Phase I/II randomized comparison	Allogeneic versus autologous BM MSCs	National Heart, Lung, and Blood Institute (NHLBI)
	TIME[b] NCT00684021 [319]	Randomized, 2 × 2 factorial, double-blind, placebo-controlled	Autologous BM mononuclear cells	National Heart, Lung, and Blood Institute (NHLBI)
	Safety study of adult MSCs to treat acute myocardial infarction NCT00114452 [262]	Randomized, double-blind, placebo-controlled, dose-escalation study	Allogeneic BM MSCs	Osiris Therapeutics, Inc.
Cancers and other disease with immune system component	Transplantation of parental haploidentical MSCs to promote engraftment in pediatric recipients of unrelated donor umbilical cord blood [320]	Phase I–II feasibility, safety and potential efficacy study	BM MSCs from the haploidentical parent donors	University of Minnesota Medical School/ Osiris Therapeutics, Inc.
	Mesenchymal stem cell infusion as prevention for graft rejection and graft-versus-host disease (GVHD) NCT00504803 [264]	Phase II, interventional, feasibility/efficacy study	BM MSCs from third-party unrelated donors	University of Liege, Belgium
	Prochymal to treat refractory acute GVHD in children undergoing allogeneic HSC transplantation [321]	Phase II, safety/ efficacy study	Allogeneic BM MSCs (Prochymal™)	Duke University Medical Center, Durham, USA
Wounds and injuries	Intravenous infusion of human adipose tissue-derived MSCs to treat a spinal cord injury [322]	Phase I, safety study	Autologous adipose-derived MSCs	RNL Bio Co, Ltd Republic of Korea
	Comparison of autologous mesenchymal stem cells and mononuclear cells on diabetic critical limb	Phase I, double-blind, randomized,	Autologous BM MSCs or BM MNCs	Third Military Medical University,

(Continued)

Table 2.1. (Continued)				
Disease Category	Example of a Trial/Trial Identifier	Trial Type	Type of MSCs Used	Sponsor
	ischemia and foot ulcer NCT00955669 [323]	controlled safety/ efficacy study		Chongqing, China
Muscle, bone, and cartilage diseases	MSC treatment of early stage osteonecrosis of the femoral head [324]	A single-center, randomized, safety/efficacy study	Autologous BM MSCs	Dalian University Hospital, China
	MSC transplantation for cartilage repair [325]	An observational safety study	Autologous BM MSCs	Osaka City University Graduate School of Medicine, Japan
Nervous system diseases	Safety and immunological effects of MSC transplantation in patients with multiple sclerosis and amyotrophic lateral sclerosis NCT00781872 [326]	Phase I/II open-safety clinical trial	Autologous BM MSCs	Hadassah-Hebrew University Hospital, Jerusalem Israel
	MSCs for the treatment of secondary progressive multiple sclerosis [327]	Phase IIa, an open-label, feasibility/safety study	Autologous BM MSCs	University of Cambridge, Cambridge, UK

^{a}The percutaneous stem cell injection delivery effects on neomyogenesis pilot study.
^{b}Timing in myocardial infarction evaluation.

responsible for joint lubrication. In the same year, Zuk et al. [57] described the derivation of MSC-like cultures from the adipose tissue; significantly the frequency of MSCs, measured as a proportion of clonogenic fibroblastic cells, was significantly higher in adipose compared to the BM, a clear advantage for the manufacture of therapeutic MSC batches (Table 2.1). Similarly, in 2003, Romanov et al. [58] described the presence of numerous MSCs in umbilical cord matrix, again their frequency in this waste product was very high [59]. MSCs were found circulating in fetal blood and also present in fetal BM and other fetal tissues [60]. Later on, similar cells were found in the placenta [61] and amniotic fluid [62, 63]; interestingly, fetal and perinatal MSCs appeared to be more proliferative than adult MSCs [59]; however, their differentiation capacities appeared to be linked to the tissue of residence [64].

In 1999, DiGirolamo et al. [65] described that prolonged MSC culture in standard conditions led to their *in vitro* ageing. Late-passage MSCs grew slower and had shorter telomeres, a clear feature of "aged" cells, and they also had noticeably lower differentiation capacity compared to early-passage cells [66]. These findings, alongside several documented reports of genetic aberrations during prolonged MSC culture [67], had a significant impact on the development of current guidelines and standards for therapeutic MSC manufacture [68, 69].

An urgent call for standardization in the MSC field was addressed by the International Society of Cell Therapy (ISCT); in 2006, the Society published a Position Statement on the Defining Criteria for MSCs [70]. Although it remains in use, in our view, it requires updating and consolidation, as discussed later in this book. According to the ISCT definition of MSCs, they are described as plastic-adherent cells, which express CD73, CD105, and CD90 antigens, lack markers of monocytes, B cells, and HSCs, and are capable of differentiation toward bone, cartilage, and fat lineages [70]. This definition is primarily based on *in vitro* expansion and differentiation assays developed by Osiris Therapeutics; positive and negative MSC markers are also mainly adopted from their Science study. Significantly, differentiation assays remain qualitative (i.e., giving no cut-off points below which differentiation is deemed absent) and no mention are given as to the degree of MSC proliferation capacity. To enumerate MSCs, Friedenstain's CFU-F assay remains in use [71, 72]; colonies of more than 50 cells (representing 5−6 cell divisions) are scored as single MSC-derived. According to more stringent criteria, true MSCs are much more proliferative; commonly 25 divisions (usually referred to as population doublings) are considered necessary to assure culture's derivation from MSCs [73, 74]; again these criteria are not included in the current ISCT definition.

2.3 MSC TISSUE HETEROGENEITY

In respect to skeletal and joint tissues, MSCs were first found in the synovium, as mentioned already [75]. Subsequently, MSCs were found in frank bone devoid of soft marrow [76], joint synovial fluid [73], joint fat pad [77], tendon [78], periosteum [74], and juvenile cartilage [79]. It has become evident that these MSCs may be responsible for physiological repair of joint structures lost as a result of "wear and tear" [80] or following injury [81]. Novel joint regeneration approaches have begun to be developed utilizing culture-expanded MSCs from the joint tissues to repair injuries to cartilage, ligament, and tendon [82, 83], not only in humans but also in animals such as race horses [84].

Pioneering studies by De Bari et al. [74, 85], who compared single-cell derived MSCs from the same donors, but derived from different tissues, have indicated that MSC differentiation propensities were dependent on their tissue of residence. Periosteal-derived MSCs were more osteogenic compared to synovium-derived MSCs [85]. At the same time, Sakaguchi et al. [86] have proven that synovial-derived MSCs were most chondrogenic compared to MSCs from other joint tissues. All cultures in these studies were compatible with the ISCT definition of MSCs; however, the latter definition was unable to capture subtle differences in their differentiation potentials, a feature critically important in selecting "right" MSCs for regeneration of a specific skeletal tissue. This further highlights a shortcoming of the current ISCT definition of MSCs—the lack of robust measures of differentiation, potentially including "molecular" potency markers in undifferentiated cells [85, 87]. Figure 2.1 illustrates this by

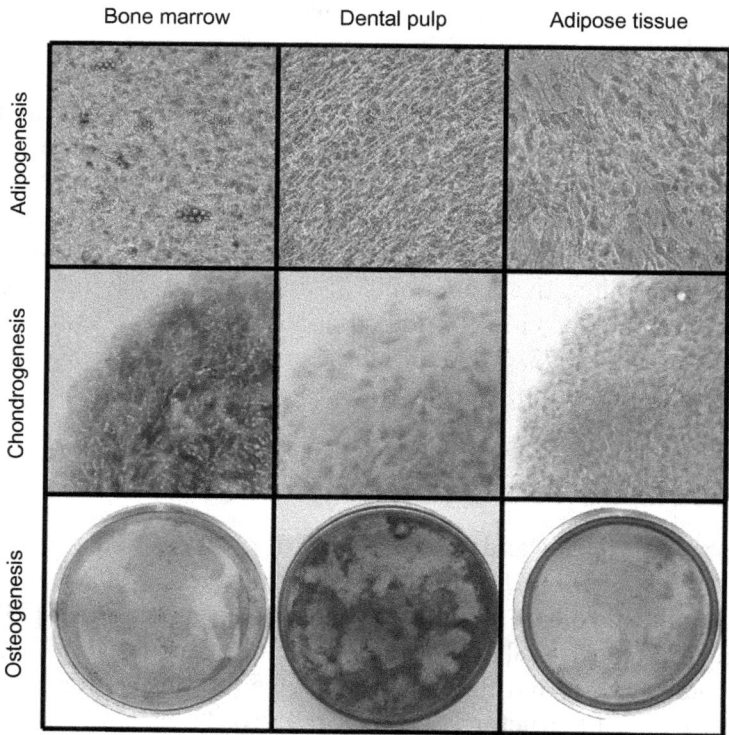

Figure 2.1 Dissimilar in vitro *differentiation capacities of MSCs expanded from human bone marrow, dental pulp, and adipose tissue (infrapatella fat pad). BM and adipose MSCs have robust trilineage differentiation, whereas dental pulp MSCs have very low adipogenesis. Adipogenesis—oil red staining, chondrogenesis—toluidine blue staining, osteogenesis—alizarin red staining on day 21 postinduction. Original magnification × 100 (adipogenesis) and × 50 (chondrogenesis). Published by Jones and Yang [171]. Elsevier.*

showing the data from our laboratory; despite an identical ISCT-compatible phenotype, MSCs from human bone marrow, joint fat pad, or dental pulp have different trilineage differentiation propensities. Dental pulp-derived and fat pad-derived MSCs are the least and the most adipogenic, respectively (Figure 2.1).

CHAPTER 3

Traditional Bone Tissue Engineering Using MSCs

3.1 PROOF-OF-PRINCIPLE EXPERIMENTS IN ANIMAL MODELS

MSCs are capable of generating bone-forming osteoblasts *in vitro*. This happens when growing MSCs are transferred from an expansion medium into the osteogenic medium [41, 88]. The latter commonly consists of chemicals known to promote osteogenesis, including a steroid such as dexamethasone, as well as a source of calcium needed for mineralization (commonly glycerophosphate) [88]. Osteoblastogenesis is normally assessed as an increased production of osteoblast-specific molecules, including alkaline phosphatise, osteocalcin, osteopontin, osteonectin, and others, followed by a mineralization stage, at which large amounts of calcium begin to be deposited [41, 88]. Notably, these *in vitro* experiments do not recapitulate the formation of osteocytes, the most mature bone-lineage cells that are trapped in the mineralized matrix of bone [89].

Initial *in vivo* experiments performed in the Friedenstein's laboratory have shown that expanded single-cell derived MSCs are capable of forming histologically mature bone in diffusion chambers [90]. The latter allow an inflow of nutrients but prevent the ingress of the host cells; by these experiments, Friedenstein has proved that newly formed bone was of the MSC origin [90]. Remarkably, this bone was formed in a nonweight bearing area and without the addition of other stimuli; on this basis, Friedenstein has initially referred to MSCs as "determined osteogenic progenitors" [47]. Subsequent animal model experiments have shown that MSCs were able to form bone following subcutaneous implantation when seeded into osteoconductive scaffolds [91]. In these "open" models, the donor origin of newly formed tissue is commonly established using species-specific probes [92].

3.2 CELL THERAPY USING MSCs

The success of these pioneering animal studies has led to the development of two types of therapeutic interventions aimed at repairing, regenerating, or replacing human bone—bone tissue engineering and

Jones: Mesenchymal Stem Cells and Skeletal Regeneration.

bone cell therapy. Generally speaking, cell therapy with MSCs commonly assumes their injection into the circulation with an aim of treating a systemic bone disease such as osteogenesis imperfecta (OI) or OP. OI is a genetic disorder, in which osteoblasts produce defective type I collagen, leading to osteopenia, fractures, and bony deformities [93]. First results of allogeneic BM transplantation in three children with OI were reported in 1999 and provided encouraging results [93]. Large-scale studies (i.e., five children) were published in 2001 [94] followed by a trial involving an additional administration of allogeneic culture-expanded MSCs (i.e., six children) [95]. This was followed by an *in utero* transplantation of allogeneic human leukocyte antigen (HLA)-mismatched male fetal MSCs into a female patient with severe OI, with encouraging initial results [96]. The rate of clinical improvement in OI children seen early after transplantation did not, however, persist for long term [97]. The most recent study from Horwitz laboratory indicates the role of nonadherent BM cells, and not only MSCs, in exerting clinical benefits [98]. No similar trials have yet been conducted for OP, most likely due to a less grave nature of the disease and the lack of an undisputed genetic link to an abnormality in MSCs.

It is noteworthy that MSC therapy can also be performed by direct MSC injection into a local bone defect, provided such defect is relatively small in volume—these approaches will be discussed later in this book.

3.3 BONE TISSUE ENGINEERING

Bone tissue engineering, as opposed to cell therapy, assumes a direct placement of MSCs, seeded on a suitable carrier or scaffold, into a defect area [99]. It is particularly pertinent for large (or segmental) bone defects that occur as a result of trauma or tumor resection. Scaffold-based approaches can also be used for maxillofacial bone regeneration, due to trauma to the head, where jaw or scull reconstruction is required [100]. The requirements to osteoconductive scaffolds have been elaborated upon at the beginning of this book. To the best of our knowledge, the first scaffold used in combination with culture-expanded BM MSCs in humans was made of macroporous bioceramics; the trial was published in 2001 [37], with 7-year follow-up results demonstrating good implant integration, no late fractures, and long-term durability of the implant [101].

The important factors to be considered, when MSCs are used in combination with scaffolds, are not only scaffold chemistry and porosity, but also scaffold surface architecture. This was recently eloquently illustrated in Dalby et al. [102] study; they showed that a novel, nanodisplaced surface topography increased MSC osteogenic differentiation, whereas highly ordered nanotopographies produced negligible differentiation. Importantly, this occurred without any chemical stimulation; the latter fact underscores the importance of mechanical stimuli including cell stretching and cytoskeletal reorientation in triggering the conversion of MSCs into osteoblasts [102].

Another key factor to take into account is a physiological and developmental "state" of seeded MSCs themselves. When MSCs were expanded in hypoxic conditions and at a low seeding density, they appeared to favor proliferation of the most immature MSCs, which can be expanded for more than 50 population doublings [103]. The mechanism behind these observations remains unclear; it is possible that standard high-density expansion in normoxic conditions leads to spontaneous, low-level MSC differentiation. An alternative explanation is that a plastic-adherent MSC population is heterogenous, as shown first by Pittenger et al. [41], and that standard MSC culture conditions do not support multipotent and bipotent clones to the same extent as that of unipotent, less proliferative clones [39].

3.4 SERUM-FREE MSC EXPANSION

A more recent development pertains to cultivation of MSCs in media lacking fetal calf serum (FCS). Animal-origin serum can be seen as a potential cause of virus and zoogen transmission to humans [104, 105]. Animal-product-free MSC media has been developed by several media manufacturers; however, its very high cost prohibits its use in large investigator-led clinical trials. Some studies have utilized human serum instead, with varying degrees of success [104, 106–108]. As early as 1995, Gronthos and Simmons [109] showed that platelet-derived growth factor-BB (PDGF-BB) and epidermal growth factor (EGF) had the greatest ability to support MSC growth. Even 10 years earlier, Hirata et al. [110] noted that PDGF was a main growth factor for CFU-F formation in human serum. It is most likely that commercial serum-free MSC culture media contains, among other ingredients, these essential growth factors. Additionally, serum contains proteins required for MSC attachment to plastic; this is why when serum-free media is used for MSC cultivation, MSC attachment "cocktails" are used in each round of passaging.

Some studies have attempted to reduce serum concentrations and supplement MSC growth media with cocktails of cytokines (PDGF-BB and EGF, grown on fibronectin [111]); others used fibroblast growth factor-2 [112] or low oxygen tension and low cell seeding density [103]. In all cases, MSC cultures have attained increased proliferative capacities (over 50 population doublings) indicating that culturing MSCs in specific culture conditions could in principle overcome inherent problems of senescence and loss of differentiation common for MSCs grown in standard FCS-containing media [103]. Interestingly, growing MSCs on an ECM from young animals was shown to "rejuvenate" MSCs from older animals, by improving their proliferation as well as osteogenic differentiation capacities [113].

MSC passaging, in general, is very laborious and hard to control. For example, level of confluency, at which MSCs are trypsinized for transfer into another flask, is normally judged by eye and is variable, even within the same laboratory. The duration of trypsinization stage is also hard to optimize because late-passage MSCs are more tightly attached to plastic, compared to early-passage MSCs. Given the above disadvantages inherent to MSC passaging, three-dimensional (3D) culture bioreactors are becoming very popular for large-scale expansion of MSCs [114, 115]. Large surface area in these bioreactors allows cells to grow and expand to a sufficient number without a need for passaging.

The most recent development in MSC cultivation is the use of platelet lyzates (PL) or platelet-rich plasma (PRP) instead of FCS [108, 116–118]. The rationale behind this technology is the fact that platelets are the most potent producers of PDGF, which is one of the factors required for MSC proliferation [109, 110]. PL and PRP are readily available from blood transfusion services and are relatively inexpensive [119]. The functionality of MSCs grown in PL-containing medium has been shown to be as good as that of MSCs grown in standard serum-containing media [117]. Furthermore, MSCs expanded in PL-containing media appeared to be cytogenetically stable and safe following transplantation to humans [68]. Table 3.1 outlines

Table 3.1. Advantages and Disadvantages of Different Media Formulations for Expanding MSCs

Types of Culture	Advantages	Disadvantages
FCS-containing media (10%)	Relatively inexpensive, good track record of safety in clinical studies	Ill-defined, batch-to-batch differences
		Potential transmission of zoogens, viruses
		Potential xenogenic immune responses
		Replicative senescence
		Loss of differentiation potential
FCS-containing media (<2%), low cell density	Increased expansion capacity, broader differentiation potential, preservation of osteogenesis	More expensive due to the use of recombinant cytokines and increased volumes of media
Serum free, with the addition of growth factors	Well-defined formulation [328, 329]	Poor growth
		The requirement for attachment factors
		More expensive [108]
Human serum	No xenogenic immune responses	Limited volume availability
		Donor-to-donor variation [107, 108]
PL and PRP	Good expansion capacity, preservation of multipotentiality [68]	Donor-to-donor variation [108, 119]

advantages and disadvantages of different media formulations for expansion of MSCs intended for bone regeneration.

Murine MSCs tend to acquire chromosome abnormalities very quickly during passaging and can form tumors following implantation [120]. Although tumor formation has been never recorded in the clinical studies published to date [68], the fact that chromosomal rearrangements might occur in human MSCs [121] should be given very serious consideration and be controlled very carefully prior to their clinical use [122, 123]. As far as PRP is concerned, it is known to be highly variable in quality depending on a donor and procedure used for its production [119]. Therefore, there is a clear need to define quality control criteria for PRP used for MSC-based therapy, potentially specifying the device used to prepare it and the time/method of storage. Of note, direct PRP injections have shown some promising results in repairing tendons and other soft tissues of the musculoskeletal system [119]. Further studies in animal models and clinical trials in humans would be needed to investigate the utility of PRP to repair bone or cartilage.

3.5 OSTEOCHONDRAL TISSUE REPAIR

Many traumatic injuries to joint involve damage to both cartilage and bone. Osteochondral grafts (i.e., those containing both tissues) are very difficult to obtain or engineer. Bone and cartilage tissues are very different in matrix composition and cell density, furthermore cartilage is avascular and aneural. Engineering of a living osteochondral constructs is therefore a considerable challenge to those working in this field. To illustrate the complexity of this challenge, it is worth mentioning a pioneering study by Prof Mao's group at the Columbia University [124]. In this study, MSCs were induced to differentiate into chondrogenic and osteogenic lineages, suspended in polyethylene glycol−based hydrogel and subsequently stratified into two separate layers, by polymerization, that were molded into the shape of a human mandibular condyle. Histological assessment demonstrated stratified layers of cartilaginous and osseous tissues [124].

CHAPTER 4

Challenges for Cartilage Regeneration

4.1 OSTEOARTHRITIS (OA) DEVELOPMENT AND THE LOSS OF CARTILAGE

OA is a degenerative joint disease that represents a growing healthcare burden, particularly in ageing Western populations [8]. Medical therapies have failed to alter the natural history of this progressive debilitating condition and the only viable definitive option remains joint replacement surgery. However, such a strategy is only feasible for a very limited number of joints including the knee and the hip and the need for new strategies has long since being recognized.

From a pathophysiological perspective, the joint destruction in OA has been historically conceptualized in terms of progressive loss of articular cartilage [125]. Indeed, end-stage OA is inevitably associated with marked attrition of cartilage with the secondary consequences being synovial inflammation, joint effusions, and subchondral bone bruising, all of which contributing to pain and ongoing joint damage. Understanding the biology of chondrocytes, the sole cellular type in cartilage and developing strategies to regenerate chondrocyte and restore their function has therefore been the center stage in joint regenerative medicine strategies in the last decades [126].

Although cartilage repair is a perfectly righteous goal, it is now clearly established that the OA process may originate in other joint structures including bone, meniscus, and ligaments [127] (Figure 4.1). In settings where disease of these other structures is the primary driver of OA, the strategies targeting solely chondrocytes may be proven a failure. It is therefore key to bear in mind that optimal OA treatment using regenerative medicine approaches should require a careful staging of the degree and type of OA and whether the cartilage repair, or repair or other joint structures including ligaments and meniscus, should be the primary focus. Nevertheless, the principles that will emerge from the knowledge of BM MSCs and joint-resident MSCs, as further discussed, have widespread applicability for regeneration of all joint components and not just cartilage.

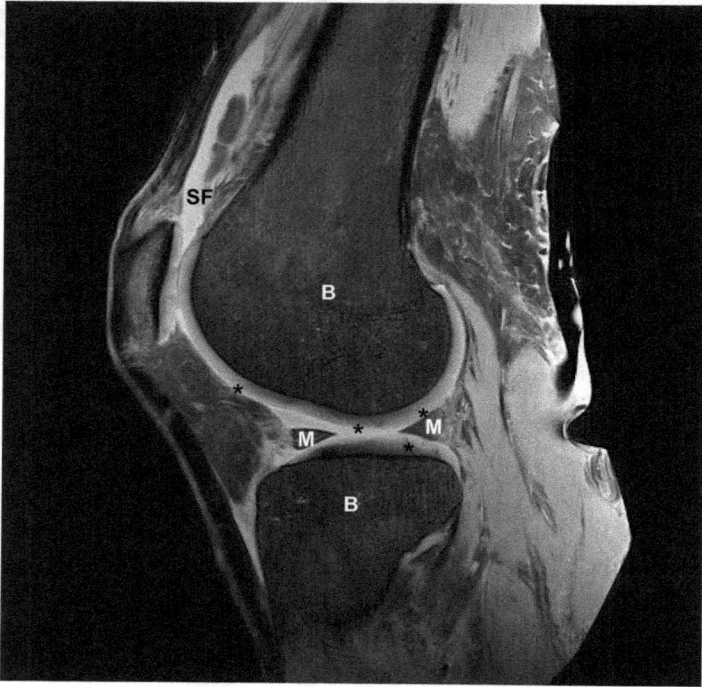

Figure 4.1 Fat-suppressed MRI of an excised cadaveric knee illustrating normal joint anatomy with cartilage (*), bone (B), menisci (M), and SF. OA may affect any tissue within the joint despite the presence of MSCs in these tissues.

4.2 CURRENT CARTILAGE REPAIR STRATEGIES

Going back over several decades, it has been noted that a modicum of cartilage repair occurs following BM stimulation techniques such as abrasion arthroplasty, drilling, and microfracture [128]. Microfracture technique involves drilling into the subchondral bone; the ensuing blood clot that formed in the defect site is believed to "mobilize" and entrap BM MSCs leading to some repair and improvement in symptoms [129]. While being associated with some temporary symptomatic improvement and the development of a scar-type fibrocartilage tissue [129], such a strategy has not been proven to be durable and is likely to work better in younger subjects and with a proper rehabilitation regime [130].

The limited repair that occurred during such procedures has led to the development of the so-called autologous chondrocyte implantation

(ACI) technique [130, 131]. In this technique, chondrocytes are pro-cured from regions of normal cartilage, culture expanded, and then reimplantated at the sites of injury [131]. A periosteal flap is commonly used to contain the repair area and to stop the egress of implanted chondrocytes into the joint fluid [131]. Over the years, several modifi-cations of this technique including the use of potency-tested [132] or matrix-assisted [133, 134] autologous chondrocytes have been attempted. Although these ACI-based methods have been used for nearly two decades, they are expensive [135] and their true cost-effectiveness remains to be assessed in the longer term (e.g., 20 years), based on the evaluation of ensuring risk of OA [136].

Besides the ACI development, the state-of-the-art pertaining to carti-lage regeneration in OA considers several factors that could be a key to better outcomes including cell selection based on MSC tissue source [137] or molecular profiling [138] and the use of novel scaffolds and membranes [126, 134, 139, 140]. As already mentioned, the majority of work in the field has utilized chondrocytes rather than MSCs as the sub-strate. This was likely based on initial demonstrations of the lack of stable chondrogenic phenotype of MSC-based constructs following their subcutaneous implantation in mice [141], as opposed to early-passage chondrocytes that produced durable cartilage [142]. BM MSCs are addi-tionally disadvantageous due to their known ability to undergo sponta-neous ossification [143]. In one study, expanded BM MSCs from OA patients were seeded onto polyglycolic acid scaffolds and differentiated to chondrocytes in the presence of parathyroid hormone-related protein (PTHrP) to regulate hypertrophy. The results showed that PTHrP inclu-sion resulted in significant suppression of type X collagen and alkaline phosphatase activity, the hallmarks of hypertrophy, without any loss of the cartilage-specific matrix proteins [143]. Other studies have utilized cocultures of MSCs with chondrocytes, showing an enhanced chondro-genesis *in vitro* [144]. Finally, another recent investigation proposed a possibility of *in situ* cartilage maturation from a fibrocartilage-type to a hyaline cartilage-type using growth factor treatment [145].

Synovium-derived MSCs appear to be more chondrogenic com-pared to fat-derived MSCs [146]. Interestingly, joint fat pad-derived MSCs possess good chondrogenesis [77, 147], possibly due to their intra-articular location and putative lubrication by synovial fluid (SF). SF MSCs that we discovered in 2004 [73] possess very good chondrogenesis at the single-cell level [148]. It is possible that this feature of SF MSCs is driven by their immediate environment in the fluid; the latter contains hyaluronic acid that has been shown to induce MSC chondrogenesis [149]. We have already alluded to the ability of low-passage cultured chondrocytes to generate durable cartilage *in vivo*; this was not, however, observed for late-passage chondrocytes that yielded fibrous tissues only [142]. In contrast, cartilage-derived MSCs, isolated using a differential adhesion assay to fibronectin [79], were shown to maintain good chondrogenicity, even after extensive expansion *in vitro* [150]; this is likely to greatly benefit the development of new cell-based therapies for repairing cartilage.

Key observations on the presence of endogenous MSCs on the cartilage surface and in the SF suggest that the use of exogenously added chondrocytes or MSCs may not actually be needed to effect joint repair, provided joint-resident MSCs can be "manipulated" to migrate into defects areas and exert the desired effect *in situ*. The observed instances of spontaneous cartilage repair outlined below suggest that these processes may occur physiologically.

4.3 SPONTANEOUS CARTILAGE REPAIR

For many years, the biological axiom has been that cartilage itself has very limited repair capacity [148]. Orthopedic surgeons have been unknowingly harnessing the power of BM MSCs for many years to effect cartilage repair through microfracture, but simultaneously held the belief that the closely juxtaposed articular cartilage itself lacks its own stem cell activity. In our opinion, this could be a result of another commonly-held view that MSC affects their repair responses via systemic circulation, whereas cartilage lacks its vascular supply. The discovery of an MSC population in cartilage superficial layer [79], in particular, argues against this historical axiom and provides a biological explanation for many incidences of spontaneous cartilage repair in human, that can occur in certain circumstances.

One example of this is osteotomy procedures, with joint realignment, which have been associated with good cartilage regeneration [151]. More recently, temporary joint unloading/distraction procedures have been shown to be effective in reducing pain and slowing down structural damage in OA [152, 153]. Both of these biomechanical approaches are based on a hypothesis that if "loading" is a major cause in development and progression of OA, then "unloading" may be able to prevent progression [152]. These biomechanical approaches may work by "mobilizing" MSCs from their dormant niches within the joint and/or by inducing their proliferation and repair responses at the defect areas.

4.4 HARNESSING THE POTENTIAL OF SYNOVIAL FLUID MSCs FOR CARTILAGE REPAIR APPLICATIONS

It is generally believed that the regenerative capacity of cartilage declines with age [154]; this can be due to the decline in the numbers of resident superficial layer MSCs. In older subjects, other joint-resident MSCs (synovium- or fat pad-derived) can in principle participate in repairing cartilage. However, since there is no direct vascular contact between synovium or joint fat and the central parts of joints where cartilage denudation occurs, the only migratory route of these MSCs can be either via the synovial lining surface or through the SF itself (Figure 4.1). MSCs can be found in healthy SF [80] and their numbers are elevated following intra-articular ligament injury [81]. The transcriptional profile of SF MSCs is similar to that of synovial MSCs [81]; this suggests that synovial MSCs can egress into the fluid as a result of either mechanical or biochemical stimulus to the synovium. Interestingly, SF MSC numbers also rise in OA [73, 80, 155, 156]; this could be a result of their increased egress from the synovium or of their enhanced proliferation in SF [80].

Novel strategies for *in situ* cartilage regeneration are therefore based on enhancing endogenous MSC trafficking and homing to the defect areas. As mentioned above, the fluid itself may contain yet unknown compounds that promote MSC egress from the synovium [157] or from the BM (latter in a case of microfracture) [158]. A comprehensive study of SF composition in different disease states will be needed to establish the nature of such compounds. Alternatively, strong gradients of chemokines, such as stromal-derived factor-1 (SDF-1), can be created artificially by embedding SDF-1 into a polymer scaffold that is placed into the defect area. Such scaffolds have been already shown to be chemotactic for synovial MSCs [159]. This "smart scaffold" approach not only "recruits" MSCs from their native niches but also directs their homing precisely into the area of interest, where their differentiation can be further enhanced by the addition of chondrogenic growth factors [160]. To harness the full potential of joint-resident MSCs, further work is needed to firmly establish the MSC "niche" in the synovium and their natural responses to joint injury, not only in animal models [161], but more importantly in human.

CHAPTER 5

Animal Models for Investigating MSC Involvement in Bone and Cartilage Repair

5.1 SMALL ANIMAL MODELS

The use of animal models in biomedical research is broadly accepted by the public; it is an essential method for preclinical testing to advance scientific knowledge and to develop new treatments and reduce suffering for both human beings and animal themselves [162]. An ideal animal model for bone regeneration should: (1) mimic clinical conditions of bone defect in order to create a permissive microenvironment that provides relevant nutrients, humidity, gaseous concentrations, and growth factors, (2) utilize fixation of the defect, as in the clinic, (3) allow the animal to apply mechanical load through the defect, (4) permit angiogenesis, (5) provide other types of cells needed for bone repair *in situ*, and (6) minimize the suffering of the animal [163].

Animal models used to study bone regeneration vary from simple subcutaneous implantation to functional complex tissue regeneration and *in vivo* bioreactors [164]. The ectopic subcutaneous implant model (Figure 5.1) is the least invasive model, broadly used for preliminary screening of new scaffold materials or MSCs from different tissue sources [164]. In this model, a small scaffold with or without MSCs or growth factors is implanted directly under the animal skin. To its advantage, this "open" model offers angiogenesis-supportive environment, which not only provides nutrient supply and gaseous exchange, but also serves as a potential route for MSC migration to the scaffold area [165]. Although a number of different species have been used for this model [166–168], the immunocompromised mouse remains the most broadly used species, as it allows the xenogeneic implantation of human cells without rejection [169, 170]. To its disadvantage, a newly formed bone may be reabsorbed with time due to lack of appropriate daily mechanical stimulation [171].

A diffusion chamber model, originally used to study angiogenesis and first used by the MSC pioneers including Alexander Friedenstein

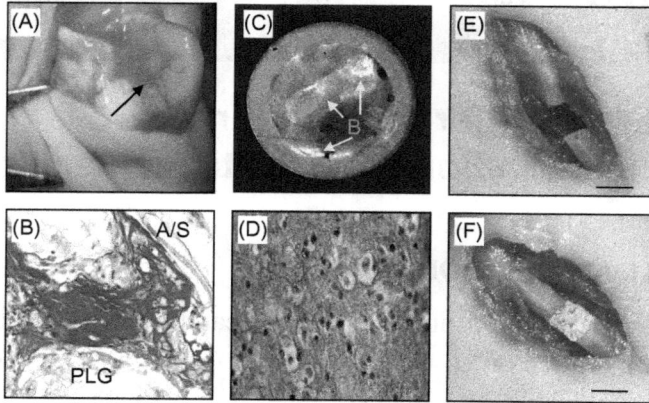

Figure 5.1 Subcutaneous implant model (A–B), diffusion chamber model (C–D), and bone defect model (E–F) in vivo. (A) Human BM stromal cells–biomaterial construct implanted subcutaneously in nude mice and vascular supply to the implant (arrow); (B) Sirius red staining showing new bone formation within the pleiotrophin absorbed poly(lactic-co-glycolic) acid (PLGA) porous scaffold; (C) X-ray images showing high-density bone nodule formation; (D) Alcian blue staining showing cartilage matrix formation within the diffusion chamber [174]; (E) segmental bone defect model in which a 2 mm bone defect created in a mouse femur; and (F) the mouse femur defect was repaired by a porous poly(lactic acid) scaffold and intramedullar pin fixation (scale bars: 2 mm). Figures (A, E, and F) are reprinted from Horner et al. [164]. Copyright @ Imperial College Press. The entire figure is published in Jones and Yang [171]. Elsevier. Figure (B) is reproduced from Yang et al. [349]. Copyright American Society from Bone and Mineral Research). Figure (D) is reprinted from Partridge et al. [174]. Copyright Elsevier.

himself [90, 172, 173], overcomes some of these limitations (Figure 5.1). This model provides a "closed" environment within a host animal to allow free exchange of nutrients (including growth factors) effectively isolating the implanted cells from the host tissues [174]. Thus, any tissue formed within the diffusion chamber must originate from the implanted cells. There are two types of diffusion chambers: with or without an injection hole. The first type is used to seal test scaffolds with or without growth factors within the diffusion chamber. At the second stage, experimental cells are injected into the diffusion chamber via an injection hole which is subsequently closed using a nylon thread. The diffusion chamber without injection hole is normally used to test cell-scaffold constructs (e.g., when cells have already grown on the scaffolds). Thereafter, the diffusion chamber is implanted intraperitoneally in a nude mouse or rat [170]. The limitation of a diffusion chamber model is the lack of angiogenic microenvironment and mechanical stimuli that are very important for bone regeneration.

Site-specific bone defect models offer distinct advantages of allowing for mechanical loading into the defect site. Many scientists refer to those models as "critical" bone defect models, but it needs to be borne in mind that the "critical" size of the bone defect (i.e., of such a size that is prohibitive of natural healing) is vastly variable between species and is additionally dependable on age, general condition, and the type of bone, even in the same species. It is also noteworthy that there are several *in vivo* bone defect models that allow for nonweight-bearing testing and repair (calvarial models) [175]. Calvarial bone defect is reconstructed using the test material in combination with cells or growth factors [176, 177]; the periosteum can be additionally used to cover the defect [178]. Calvarial model can utilize several small species (e.g., mouse, rat, or rabbit) [179, 180].

For weight-bearing testing, the defect can be made in long bones (e.g., the femur or tibia) [181] (Figure 5.1). The most commonly used methods for creating a critical weight-bearing defect are either an osteotomy or a traumatic approach. Osteotomy surgically removes the required length of bone, producing a consistent defect with a "clean" cut. A traumatic injury produces a jagged-edge bone defect, including the trauma to surrounding soft tissue, which better reflects the real conditions of fracture in humans. This is achieved using a three-point bending device or an impact device [182, 183]. Following the creation of the defect, it subsequently reconstructed with the test material alone or in combination with cells and growth factors, and fixed externally [184, 185] or internally [186].

5.2 LARGE ANIMAL MODELS FOR BONE REPAIR AND REGENERATION

As mentioned above, small animals such as nude mouse or rat are very useful for testing human cells *in vivo*. However, considering the body mass and defect sizes, models in large animals (e.g., sheep, dog, pig, and goat) are additionally required to reflect the real clinical situations [164, 187]. To date, different large animal species have been used to model fracture healing and to test novel tissue engineering approaches. As early as 1998, Bruder et al. [188] reported to use autologous BM MSCs and porous ceramic scaffolds to repair a 21 mm osteoperiosteal segmental cortical defect in an adult dog. After 16 weeks, radiographic observation showed a large osseous callus formation in scaffold plus MSCs group compared to the controls. Similarly, Kon et al. [189] used a hydroxapatite cylinder loaded with

BM MSCs to heal a 35 mm sheep tibia bone defect. At 8 weeks, more callus was observed in animals receiving cell-loaded implants compared to controls, which were confirmed by radiographical and histological assessments as well as by mechanical testing.

Apart from sheep models [190–193], "critical size" defects have been created in goats [194]. In the study by Xu et al. [194], 30 mm diaphyseal femoral defects were filled with allogenic DBM and autologous BM MSCs. Radiological analysis and biomechanical evaluation were performed at 12 and 24 weeks after the operation and X-ray examination showed excellent bone healing in the DBM-MSC group, whereas only the DBM group was characterized with only partial bone repair; no healing was observed in untreated controls. As mentioned above, calvarial defect models are normally performed in smaller animals. In one recent study, however, Kinsella et al. [195] reported the use of absorbable collagen sponge to deliver BMP-2 with corticocancellous chips to treat calvarial defects in 12–13-months-old beagles. Both rediopacity and histology showed consistent ectopic bone formation in all treatment groups.

5.3 ANIMAL MODELS FOR OSTEOCHONDRAL TISSUE ENGINEERING

In the so-called osteochondral defects, damaged areas span both the articular cartilage and the underlying subchondral bone. In order to repair an osteochondral defect, it is crucial to consider not only the bone and cartilage tissues *per se*, but also the bone–cartilage interface. In the past decade, tissue engineering has emerged as a potential solution to tackle all three components [196] and considering the complexity of the osteochondral tissue engineering, it is essential to test the chosen strategies *in vivo*. Although osteochondral defect can be created in small animal models (e.g., rabbits and rodents), it is difficult to create such a surgical model that is analogous to the clinical situation. In comparison to the mouse and the rat, the rabbit is relatively large in body weight and articular cartilage surface area. The osteochondral defect is most commonly made in rabbit patella groove [197–199]. Recently, Nishino et al. [200] carried out resection of the articular cartilage and subchondral bone from the entire tibial plateau of 16 rabbits. Their findings suggested that weight bearing had a positive effect on the quality of the regenerated cartilage. Large animal models that were used to test osteochondral repair include dogs [201], goats [202, 203], and sheep [204, 205].

In 2010, Ho et al. [206] reported the creation of critically sized osteochondral defects at the medial condyle and patellar groove of pigs. Autologous MSCs were seeded, via fibrin, onto a biphasic implant comprising of a polycaprolactone (PCL) cartilage scaffold and a PCL–tricalcium phosphate osseous matrix. The defect was filled with the construct and resurfaced with a collagen mesh, which served as a substitute for periosteal flap in preventing cell leakage. Similarly, Im et al. [207] used a porcine model to test the efficacy of a biphasic scaffold for the repair of osteochondral defects.

5.4 ANIMAL MODELS FOR CARTILAGE TISSUE ENGINEERING

An *in vivo* cartilage defect model needs to mimic partial-thickness defects to cartilage, that is, those that do not penetrate into the subchondral bone. Therefore, they should provide a microenvironment that lacks angiogenesis and undergoes mechanical stimulation appropriate for chondrogenesis and regeneration of articular cartilage. The anatomical location of the defect must be carefully considered in respect of whether the site is to be weight bearing or nonweight bearing, as well as the size and type of the defect according to the experimental design and the animal species being used. The use of animal models for testing the efficacy of cartilage repair is limited due to the anatomical differences of human and animal cartilage. According to Hunziker and coworkers [208, 209], cartilage thickness and overall cell volume densities vary greatly across the different species. In 2010, Shimomura et al. [210] used synovial MSCs to develop 3D tissue-engineered constructs which were used to repair chondral defects (8.5 mm diameter and 2.0 mm depth, which did not breach the subchondral bone) in the medial femoral condyle of both immature and mature pigs. Interestingly, tissue-engineered constructs promoted the repair of a chondral lesion in both immature and mature pigs; the repaired tissue also exhibited viscoelastic properties similar to normal cartilage regardless of the skeletal maturity.

Another issue to consider is that, damage to cartilage is usually preceded by chronic degenerative disease (e.g., OA). The presence of chronic degeneration with multiple underlying causes is neglected in most animal models and yet represents the largest challenge to any reparative attempt [164]. In 2008, Gelse et al. [211] reported the use of a porcine OA animal model to investigate the potential of using transgene-activated periosteal cells for permanent resurfacing of large partial-thickness cartilage defects. The results indicated that such defects can be resurfaced efficiently with hyaline-like cartilage; however, the long-term stability depended on biochemical factors that were active only in deeper zones of the cartilaginous tissue.

CHAPTER 6

Native MSCs

6.1 MSCs AND MARROW RETICULAR CELLS

As mentioned in the previous chapters, ISCT's position statement defines MSCs retrospectively, that is, following culture expansion in standard conditions. It does not provide any real indication as to what MSCs look like in their *in vivo* niches. To address these questions, the surface phenotype of native MSCs in different tissues needs to be defined. A range of assays and reagents required to address this issue should include: (1) a minimum panel of candidate markers potentially specific for MSCs (ideally including "negative" markers/controls); (2) sorting candidate "MSC" and control "non-MSCs" populations from the same tissue; (3) their propagation in culture in order to establish relative proliferative capacities and finally; (4) differentiation assays to test for multipotentiality. Most robust investigations also included the study of clonogenicity at the single-cell level (using limiting dilution assays) [212], single-cell PCR [213], and functional testing of transplantability and self-renewal capacity in an animal model [34]. In many studies, native MSC morphology [40, 212, 214] and topography [213, 215] were also demonstrated using cytochemical and immunohistochemical assays.

Most comprehensively, this inquiry has been undertaken for human BM MSCs. Following a decade of "searching" and testing individual candidate markers (Table 6.1), a consensus is now emerging regarding the *in vivo* phenotype and topography of MSCs in human BM. In a mouse system, this search seems to be even more complicated; this is likely due to relatively large variability between different mouse strains [216, 217] and the above-mentioned problems with long-term MSC cultivation [120]. Surprisingly, the most definitive up-to-date marker of mouse BM MSCs was identified in the study of circadian rhythms of mouse HSC release into the systemic circulation [218]. This was shown to occur with the help of nestin-positive marrow stromal cells that had all the characteristics of MSCs [218].

Table 6.1. Positive Markers for the Identification and Purification of Human BM MSCs (From Year 2000 Onward)

Marker	Year	Reference
CD105	2000	[330]
CD271	2002	[331]
D7-FIB, CD271	2002	[40]
Stro-1/CD106	2003	[212]
CD49a	2003	[332]
Stro-1/CD146	2003	[250]
D7-FIB, CD105	2003	[333]
CD73, CD105, CD90	2005	[334]
CD105	2006	[335]
ALP[a]	2007	[336]
D7-FIB, CD106	2007	[281]
GD2	2007	[337]
SSEA-4	2007	[338]
CD73	2007	[339]
CD271, CD140b, CD340, CD349	2007	[340]
CD146	2007	[34]
CD105	2008	[341]
CD200	2008	[342]
CD146	2008	[343]
Fibroblast activation protein α	2008	[344]
PODXL	2009	[333]
Stro-4	2009	[345]
ALP[a]	2010	[346]
CD271	2010	[347]
CD271/CD146	2011	[213]
CD271/CD146	2012	[348]
[a]*Nontissue-specific alkaline phosphatase.*		

Many "*in vivo*" MSC markers including Stro-1 and CD271 are lost during standard MSC cultivation [40, 50, 214]. It can be hypothesized that their expression on MSCs *in vivo* is a result of MSC interactions with neighboring cells and ECM proteins (commonly referred to as stem cell "niche"). Not surprisingly, these features are difficult to reca-pitulate in 2D standard cultivation systems. Nestin-positive MSCs have been shown to closely interact with vasculature and nerve fibers in a mouse BM [215]. CD271-positive human BM MSCs have a similar topography of reticular cells that form an interconnecting network with vasculature and extend to bone surfaces [40, 213, 219, 220]. Both mouse nestin-positive MSCs and human CD271-positive native MSCs produce large amounts of a cytokine CXCL12/SDF-1, as well as bone-related proteins (osteocalcin, osteopontin, and osteonectin) and fat-related proteins [215, 221]; this highlights a fundamental similarity in human and murine BM MSCs *in vivo* despite different markers used for their purification.

6.2 NONADHERENT MSCs

It is worth noting the fact that all MSCs should be necessarily adherent is not yet unequivocally proven [222]. The fact that MSCs can be grown in a liquid culture, with good viability and preserved functionality, was shown in 2003 [223]. MSCs can be frequently found in circulation of many animal species and less commonly in humans [30]. Viable cells can also be recovered from other body liquids such as synovial [73] and amniotic [73] fluids. Recent trials for OI further supported a potential role of nonadherent MSCs in bone regeneration.

Our recent study in which we compared the frequencies of CD271-positive cells and CFU-Fs from the same donors [224], as well as several other independent studies that used markers other than CD271 to sort for native MSCs [212, 213], have indicated that only a proportion of sorted cells have actually produced CFU-Fs (between 1% and 20%). This suggests the possibility of a nonadherent MSC subpopulation that is immediately lost during standard MSC cultivation and CFU-F assay. This idea can be further supported by the observed dramatic differences in the MSC transcriptome between freshly isolated and culture-expanded MSCs [221, 225]. Different patterns of expression of integrins and other matrix-binding molecules in particular [225] strongly suggest a potential loss of as yet unknown, nonadherent MSC subpopulation. Further study in this field is needed to identify the phenotype of such putative subpopulation and its physiological role *in vivo*.

6.3 SOLID TISSUES: MSCs AND FIBROBLASTS

In vivo MSC identification in solid connective tissues is additionally complicated by the presence of fibroblasts, common mesenchymal-lineage cells present in most solid tissues. The "borderline" between fibroblasts and MSCs has always been rather obscure [226]. Skin fibroblasts, the most studied fibroblastic cell type, express all ISCT-approved MSC markers [73, 227, 228]; they also show a degree of multipotentility [229–231] and immunoregulatory capacity [232]. Similar to MSCs, skin fibroblasts are quite proliferative, particularly in young donors, with some clones able to propagate beyond 20 population doublings [233], a cut-off point commonly used for MSCs. Furthermore, fibroblasts display "a positional memory," that is, different transcriptional and surface marker profiles depending on the part of the body they are derived from [233, 234]. This is reminiscent of the observed "tissue-specific" profiles of MSCs resident in anatomically different tissues [85, 235]. Therefore, in our opinion, a conceptual framework allowing a clear discrimination of an MSC from a mundane fibroblast has not been yet constructed. On this basis, the identification of "positive" MSC markers in solid connective tissues such as adipose [236], placenta [237], or synovium [238] remains rather empiric. Some candidate markers have been derived from high-throughput screening by quantitative Polymerase Chain Reaction (qPCR) and flow cytometry [227, 228]; however, no consensus yet exists on the native phenotype of MSCs in the majority of connective tissues.

In relation to bone itself, Noth et al. [239] and Tuli et al. [76] were the first to show that it harbors large numbers of MSCs. They utilized explant cultures showing MSCs "outgrowing" from fragments of bone, even when enzymatically "cleaned" of soft marrow. In 2005, Sakaguchi et al. [240] demonstrated that bone-resident MSCs are present in high numbers in enzymatically digested cellular fractions [240]. In 2010 and 2012, our studies showed that bone-resident MSCs, both in trabecular [241] and cortical bone [242], have the CD271-positive phenotype. These bone-resident MSCs are very similar in functionality to BM MSCs, possibly due to their intraosseous location [241, 242]. Relative functionalities of CD271-positive MSCs that are attached to bone and those located in interstitial space, in a guise of marrow reticular cells [213, 220], are yet to be established.

6.4 MSC AS A PERICYTE

Recently, a revolutionary concept of an MSC as a pericyte has been popularized by a pioneer MSC, biologist Arnold Caplan [243]. This concept postulates that in all tissues, MSCs have the topography of pericytes [243–245]. Pericytes are defined topographically as contractile cells specifically located surrounding the endothelial cells and supporting blood vessel integrity [246]. The concept is based primarily, but not exclusively, on recent studies by Sacchetti et al. [34] and Crisan et al. [244]. However, it is worth mentioning that this concept has a somewhat longer history. In 1998, a pioneering study by Doherty et al. [247] demonstrated that vascular pericytes have an inherent osteogenic potential. In 2001, Bianco et al. [248] proposed the pericyte identity of MSCs in the marrow. In 2003, Short et al. [249] as well as Shi and Gronthos [250] gave their support to this hypothesis; however, only recently has strong and independent experimental evidence using the CD146 marker been obtained to substantiate these ideas and to prove a perivascular topography of MSCs in a number of tissues [34, 244]. These results should, however, be taken with a certain degree of caution. Besides the pericytes, CD146 is also expressed on endothelial cells [251, 252] that are located in close proximity to pericytes. CD146-based MSC isolation approaches using solid tissue digests are therefore bound to coisolate resident endothelial cells leading to mixed cultures, at least early in cultivation. Furthermore, the authors of this concept acknowledge that not all pericytes are MSCs [253] and the markers differentiating "MSC" from "non-MSC" pericytes are yet to be found.

A pericyte concept for MSC topography is very appealing; it explains the abundance of MSCs in solid tissues, given their almost ubiquitous vascular supply [254, 255]. Furthermore, in their recent review, Caplan and Correa [256] propose a novel concept for MSC "mobilization" approaches by using cytokines and other factors to "unlock" MSCs from their pericytic niches thus allowing their enhanced migration through the tissue and toward the damaged areas. Although as yet hypothetical, this concept is very valuable in terms of developing novel approaches for skeletal repair based on native MSCs, the issue that will be discussed in the last chapter of this book. To add further complexity to this very appealing theory, some recent cell-tracking studies have shown that a proportion of MSCs in some tissues do not have a topography of pericytes [161, 257]. Another, extreme example of this is cartilage, a completely avascular tissue, in which MSC activity was found to be present in its superficial layer [79, 150].

Based on all of these studies, it is possible to speculate that MSC activity may be inherent to a plethora of mesenchymal-lineage, connective tissue cells, including some fibroblasts and pericytes, which may lay dormant but can be rapidly activated in response to external stimuli. The MSCs "fates" may not be limited to differentiation and can also include migration, homing, growth factor release [258] and, as proposed recently, MSCs are "polyrized" towards one of these two phenotypes [259]. All of these responses are likely to contribute to the regenerative capacity of MSCs *in vivo*.

CHAPTER 7

"Trophic" Actions of MSCs

7.1 DISCOVERY OF THE TROPHIC ACTION OF MSCs IN CARDIAC REPAIR TRIALS

As mentioned in the previous chapters, the rationale for using MSC for repairing bone has originated from Friedenstain's diffusion chamber experiments, in which expanded BM MSCs were able to form histologically normal bone via their differentiation to osteoblasts [90]. According to the Caplan's model [49] and supported by several initial experiments using mesenchymal cell lines [260], MSCs were similarly shown to differentiate into muscle myoblasts [261]; this has formed the foundation for several clinical trials, chiefly supported by Osiris Therapeutics, for the treatment of myocardial infarction using MSCs [262]. However, ensuing animal model experiments and clinical trials data have unexpectedly shown very minimal differentiation of MSCs to myoblasts [263]; furthermore, MSCs were very short-lived *in situ* and appeared to exert their action via a release of factors supporting the survival and improving the functionality of resident cardiac progenitor cells [263].

A similar "trophic" mechanism of action is believed to occur with MSCs used for graft-versus-host disease (GVHD) treatment [264, 265]. The mechanism of immunosuppressive actions of MSCs in this disease setting is not yet completely understood [266], but it seems to be based on the MSC immunoregulatory action, which is mediated via a release of numerous cytokines and metabolites such as indoleamine 2,3-dioxygenase and prostaglandin E2 [267]. Additionally, MSCs are known to trigger cellular cascades of events leading to "dampening" immune responses via the functions of many immune cell types, including T cells, B cells, natural killer cells, monocyte/macrophages, dendritic cells, and neutrophils [268, 269].

7.2 OTHER DISEASE APPLICATIONS

In relation to neurological disease applications such as spinal cord injury [270], multiple sclerosis [271], or amyotrophic lateral sclerosis

(ALS) [272], MSC "neuroprotective" action is worth mentioning. The growth factors implicated in MSC neuroprotection include brain-derived neurotrophic factor (BDNF), nerve growth factor (NGF), neurotrophin-3 (NT-3), glial cell-derived neurotrophic factor (GDNF), and some members of basic fibroblast growth factor (FGF) family [270]. Interestingly, MSCs have also been shown to inhibit immune-mediated damage to neurons and rescue them from apoptosis; the molecular mechanisms for this are yet poorly understood however the inhibition of oxidative stress molecules by activated macrophages and microglia known to damage neurons have been implicated [271, 273]. In relation to liver regeneration applications, the secretion of growth factors such as hepatocyte growth factor (HGF) by infused MSCs is believed to trigger paracrine mechanisms that promote the survival and proliferation of hepatocyte progenitors and other liver-resident regenerative cells such as oval cells and stellate cells [274, 275].

Haemopoiesis-supportive growth factor production by cultured BM MSCs has been first demonstrated in a study by Majumdar et al. [276]. MSCs were shown to constitutively express mRNA for interleukin (IL)-6, IL-11, leukemia inhibitory factor (LIF), macrophage colony-stimulating factor (M-CSF), and stem cell factor (SCF). More recently, our group have shown that the amount of produced CXCL12/SDF-1 by native uncultured BM MSCs is $\sim$1,000-fold higher compared to cultured MSCs [221]. SDF-1 is a potent chemoattractant for both HSCs [277] and MSCs [278, 279].

The need for direct interactions between MSCs, HSCs, and neurons to fine-tune local SDF-1 production by MSCs has been recently shown in a murine system [215, 218] leading to a novel concept of the pivotal role of the SDF-1/CXCR4 axis in determining HSC fate *in vivo* [280]. This cross-talk is certainly lacking in conventional 2D MSC cultures potentially leading to the down-regulation of their innate SDF-1 production [221]. Native BM MSCs can also produce *B-cell* activating factor (BAFF), the cytokine needed for maturation and survival of B cells in the BM [281]. Interestingly, MSCs not only produce soluble ligands such as SDF-1 and IL-7 [215, 282], but they also express surface receptors for these molecules [282, 283]; this points toward autocrine mechanisms of regulation of the respective molecules.

CHAPTER 8

Novel Approaches for Bone Regeneration Targeting Native MSCs

A regulatory path for cell therapies based on culture-expanded MSCs is very long and the cost of one "dose" of cell therapy treatment remains very high (considerably higher for autologous compared to allogeneic MSCs) [135]. Not surprisingly, scientists, clinicians, and biotech companies alike seek new ways to develop treatments that involve shorter regulatory routes and lower manufacturing costs. Examples of such solutions based on native, minimally manipulated MSCs are outlined below.

8.1 MEDICAL DEVICES ALLOWING MSC CONCENTRATION

Business analysts refer to this approach as a "new on-site solution"; it takes advantage of no transportation costs and the use of fairly simple equipment that can be housed at the hospital site. In these settings, native uncultured MSCs are obtained using cell separation/centrifugation devices and cells are implanted immediately using the same day surgery. Such devices are commonly based on unique physical characteristics of MSCs such as their size or cellular density. The concept of MSC filtration based on their unique size was first proposed in Hung et al. [284]. For MSCs residing in the BM, centrifugation devices can be also based on differential cell density and such devices have already been tested in both animal [285] and human studies [286−289]. These devices primarily eliminate red cells and concentrate MSCs in a smaller volume [290] allowing their direct injection or implantation (in conjunction with a carrier) into a defect. Of note, large volumes of marrow are required to achieve sufficient concentration of MSCs per unit volume [291]. Additionally, MSCs are only "concentrated" but not "isolated," which means that the injected cell mixtures contain other, potentially inhibitory, cells. Nevertheless, concentrated cellular preparations containing native BM MSCs have been already used in many clinical studies with reported safety [289] and, in many cases, good efficacy in repairing nonunion fractures [291], osteonecrosis of femoral head [292], and bone cysts [293].

Similar devices have been developed for fresh MSC extraction from subcutaneous fat/lipoaspirated tissue [294], another readily available source of MSCs. Similar to BM "concentrators," native adipose MSC preparation involves centrifugation; however, a brief tissue digestion step using a low concentration of collagenase is sometimes used to allow better tissue dissociation [294, 295].

8.2 HOST MSC "MODIFIERS"

These are small molecules that, once administered, control host MSC functionality [135]. The downstream effects can be broadly divided into: (1) molecular control of MSC differentiation toward the desired lineage and (2) directing host MSC migration toward the repair site. There are growth factors other than BMP-2 and -7 that have been so far tested for their potential to regenerate bone [296]. These include PDGF, vascular endothelial growth factor (VEGF), and FGF [26]. PDGF is manufactured by Biomimetic Therapeutics and is available in a recombinant form for the treatment of periodontal bone defects [26]. It does not induce MSC differentiation but is believed to have a mitogenic effect and enhances MSC stimulation of angiogenesis [26]. Recombinant human FGF-2 (rhFGF-2) in a gelatin hydrogel form has been used in a prospective, randomized, placebo-controlled clinical trial of 70 patients, showing an accelerated healing of tibial shaft fractures with a good safety profile and no adverse events [297]. To date, no published clinical studies utilizing VEGF alone or in combination with other growth factors are available. A systemic administration of new growth factor, insulin-like growth factor-I (IGF-1) in conjunction with cultured MSCs, was recently evaluated in a mouse model [298]. The increase in soft and new bone tissue volumes, which correlated with increased biomechanical toughness, was documented [298].

As mentioned earlier in this book, PDGF, produced chiefly by platelets, is additionally believed to unlock MSCs from their native pericyte niches allowing their free migration [256]. PDGF is present in platelet-rich concentrates (PRPs) and some clinicians use autologous PRP instead of purified recombinant growth factors for augmenting musculoskeletal tissue repair [119]. The advantages of PRP include low manufacturing costs and the presence of other potentially useful growth factors. More promising results have been observed, however, in relation to soft rather than hard tissue repair. Current lack of consistency in results emanated from various PRP-based clinical studies can be explained by the absence of standard protocols for the production and characterization of PRP (e.g., in relation to platelet/leukocyte ratios) [119].

The second group of host MSC modifiers includes molecules enhancing MSC migration. Most interesting of those is a chemokine SDF-1, whose role in MSC/HSC interaction was already outlined in previous sections. SDF-1 chemotactic gradients have been shown to influence MSC migration in animal studies [278, 299]. Most recently, SDF-1 was shown to stimulate bone growth by additionally mediating chondrocyte hypertrophy [300]. In relation to cartilage repair, both SDF-1 and transforming growth factor beta (TGF-beta) have been shown to act in concert attracting synovial MSCs to the sites of cartilage damage and inducing their chondrogenesis [159]. Recently, type 1 collagen scaffold containing SDF-1 was used in a rabbit model to create a matrix environment conducive to synovial MSC migration and retention [301]. This provided proof-of-principle that partial-thickness cartilage defects can be also repaired [301] by using smart scaffolds triggering endogenous MSC homing [160].

8.3 "IN VIVO" BIOREACTORS AND PERIOSTEUM-LIKE MEMBRANES

These are used primarily for the repair of large maxillofacial or segmental bone defects. The example of a former is a pioneering study of Warnke et al. who were the first to demonstrate the feasibility of treating an extended mandibular discontinuity defect using a "living bioreactor" [100]. Three-dimensional computed tomography was used to produce a titanium mesh cage that was filled with bone mineral blocks and infiltrated with BMP-7 and autologous BM. The transplant was implanted into the latissimus dorsi muscle and 7 weeks later transplanted as a free bone—muscle flap to repair the mandibular defect. Six months after transplantation, patient's speech and mood improved and bone formation was detected in all parts of the mandible replacement [302]. The patient enjoyed improvement in his quality of life, but unfortunately died as a result of cardiac arrest 15 months after implantation [302]. Although initially successful, such complex tissue-engineering procedures remain to be validated in large-scale studies before they can be used routinely in clinic [303].

The second example is a novel approach for the reconstruction of critical size long bone defects called the Masquelet technique [304] . The reconstruction requires a two-stage approach; at the first operation radical debridement is undertaken and cement spacer is implanted at the site of bone defect. Significantly, the cement spacer not only acts as a mechanical stabilizer and a barrier to stop fibrous tissue invasion, but it also represents a foreign body, inducing the formation of vascularized "pseudomembrane" around the spacer [305]. At the second stage, approximately 6−8 weeks later, the cement spacer is carefully removed and the defect is filled with autograft material or its substitutes. The induced membrane acts as a "biological chamber" and possesses osteoinductive, osteogenic, and angiogenic properties [306]. The advantage of the induced membrane is that it contains the bone graft and prevents its resorption; additionally it is rich for growth factors implicated in vascularization and osteogenic differentiation [305, 306]. An increased production of VEGF and BMP-2 by the membrane suggests the presence of membrane-resident MSCs, which are likely to have migrated from the surrounding tissues.

Current efforts are directed toward fabricating a similar, but "artificial" membrane; this would reduce the treatment to a one-stage procedure. In one recent study, a permeable collagen membrane/wrap was used in conjunction with a hydroxyapatite bone graft to repair 10-mm segmental long-bone defect in rabbits [307]. The use of the collagen wrap showed increased bone ingrowth and periosteal remodeling. The authors proposed that, similar to "induced" membrane in Masquelet technique, the wrap contains the local bone-healing environment while reducing fibrous infiltration. In another recent study, a periosteal substitute was engineered using a PRP membrane incorporating autologous BM MSCs and wrapped around an osteoconductive scaffold for regeneration of segmental bone defect in a rabbit model. The results provided good evidence of membrane's capacity to biomimic a periosteal response and to enhance bone regeneration [308].

8.4 TARGETING MSCs IN SYSTEMIC BONE DISEASES: NOVEL BONE ANABOLICS

Compared to fractures, systemic bone diseases including OI and OP are significantly harder to treat with MSC injections. To have a systemic effect, an effective treatment should be delivered via the systemic circulation, but this is known to lead to MSC loss due to their entrapment in lungs [309]. One recent study has proposed that injecting native uncultured MSCs, in comparison to cultured MSCs, may lead to their lesser retention in lungs due to a different integrin-expression pattern [225]; proof-of-principle experiments in animal models are needed to test this idea. Intra-bone injections have been attempted in animal models and resulted in better MSC engraftment into the marrow [310]. The pioneers of this concept believe that the injection of cells directly into the BM cavity facilitates the engraftment of both donor HSCs and MSCs and that using this method many age-associated bone diseases, including OP, could be potentially treated [311].

At present, several "anabolic" (i.e., targeting the osteoblast lineage) approaches for OP treatment are being developed based on the modulating of Wnt pathway activity in native MSCs [312]. These treatments fall into a category of "host MSC modifiers." OP MSCs have been recently shown to have an enhanced mRNA expression of genes coding for inhibitors of Wnt signaling, such as sclerostin [313]. Wnt proteins are extracellular glycoproteins that can activate an intracellular signaling leading to an accumulation of a protein called beta-catenin. Wnts bind to the membrane receptors frizzled and lipoprotein-receptor-related protein 5/6 (LRP5/6). In the absence of Wnt, beta-catenin is degraded via the ubiquitin−proteosome pathway. In the presence of Wnt, the protein complex is disrupted and beta-catenin translocates to the cell nucleus and binds to transcription factors that affect the expression of Wnt-responsive genes, which are important in bone formation [312]. Inhibitors of Wnt signaling can bind to frizzled (serum frizzled-related proteins), Wnts or LRP5/6 (sclerostin and dickkopf-1). These inhibitors prevent Wnt from activating the signaling pathway, leading to a decrease in signaling and a consequent reduction in bone formation. High sclerostin expression in OP MSCs can therefore explain their poorer osteoblastogenesis [313]. Antibodies to sclerostin and another Wnt inhibitor, dickkopf-1, have been shown to stimulate bone formation in animal models leading to ongoing clinical trials aimed to evaluate their effects in humans [314].

Another approach involved further activation of Wnt signaling in OP MSCs by inhibiting glycogen synthase kinase 3 (GSK-3), an enzyme that phosphorylates beta-catenin, making it amendable for degradation via the ubiquitin−proteosome pathway [312]. In one recent study, a novel inhibitor of GSK-3, AR28, was used in a BALB/c mouse model demonstrating an enhanced bone mass after 14 days of treatment [315]. The authors concluded that the increased bone mass was the result of early amplification of bipotent MSC clones (osteogenic and adipogenic), which was driven to osteoblast differentiation at the expense of adipogenesis. Therefore, GSK-3-beta can represent an attractive therapeutic target not only for cancer chemotherapy [316], or to treat inflammation [317], but also for inducing osteoblastogenesis and regenerating bone in OP [315].

CONCLUDING REMARKS

Regenerative medicine approaches offer new and exciting possibilities for repairing bone, cartilage, and other skeletal tissues. The choice of available "cellular" options is growing exponentially, with new types of MSCs being discovered and characterized almost on a monthly basis. Similarly, the biotech industry has been very active in developing new scaffolds and biomaterials capable of delivering these cells into the defects. Animal models have been used very extensively to "bridge the gap" between blue-sky, proof-of-principle *in vitro* studies, and clinical trials in humans and such works should continue. Importantly, economic "viability" of new approaches and appropriate business and adoption models should be considered from the start, to ensure the uptake of these new therapies by health care providers in different countries. An effective dialog between scientists, clinicians, business professionals, and regulators is therefore of a paramount importance to enable these new MSC-based therapies to become a clinical reality in the twenty-first century.

REFERENCES

[1] Bielby RC, Jones EA, McGonagle DG. The role of mesenchymal stem cells in maintenance and repair of bone. Injury 2007;(S1):S26−32.

[2] Crockett JC, et al. Bone remodelling at a glance. J Cell Sci 2011;124(7):991−8.

[3] Calori GM, et al. Risk factors contributing to fracture non-unions. Injury—Int J Care Injured 2007;38:S11−8.

[4] Dimitriou R, et al. Bone regeneration: current concepts and future directions. Bmc Med 2011;9.

[5] Jones KB, et al. Cell-based therapies for osteonecrosis of the femoral head. Biol Blood Marrow Transpl 2008;14(10):1081−7.

[6] Jones EA, et al. Mesenchymal stem cells and their future in osteoporotic fracture repair. Adv Osteoporotic Fract Manage 2006;4(3):11−6.

[7] Sahota O, Morgan N, Moran CG. The direct cost of acute hip fracture care in care home residents in the UK. Osteoporosis Int 2012;23(3):917−20.

[8] Arden N, Nevitt MC. Osteoarthritis: epidemiology. Best Pract Res Clin Rheumatol 2006; 20(1):3−25.

[9] Corti MC, Rigon C. Epidemiology of osteoarthritis: prevalence, risk factors and functional impact. Aging Clin Exp Res 2003;15(5):359−63.

[10] Dagenais S, Garbedian S, Wai EK. Systematic review of the prevalence of radiographic primary hip osteoarthritis. Clin Orthop Relat Res 2009;467(3):623−37.

[11] Sofat N, et al. Recent clinical evidence for the treatment of osteoarthritis: what we have learned. Rev Recent Clin Trials 2011;6(2):114−26.

[12] Baron R, Hesse E. Update on bone anabolics in osteoporosis treatment: rationale, current status, and perspectives. J Clin Endocrinol Metab 2012;97(2):311−25.

[13] Drosse I, et al. Tissue engineering for bone defect heating: an update on a multi-component approach. Injury—Int J Care Injured 2008;39:S9−20.

[14] Giannoudis PV, Einhorn TA, Marsh D. Fracture heating: the diamond concept. Injury—Int J Care Injured 2007;38:S3−6.

[15] Spiegelberg B, et al. Ilizarov principles or deformity correction. Ann Royal Coll Surg Engl 2010;92(2):101−5.

[16] Malizos KN, et al. Free vascularized fibular graft—a versatile graft for reconstruction of large skeletal defects and revascularization of necrotic bone. Microsurgery 1992; 13(4):182−7.

[17] Cox G, et al. Reamer−irrigator−aspirator indications and clinical results: a systematic review. Int Orthop 2011;35(7):951−6.

[18] Bohm P, Renner E, Rossak K. Massive proximal femoral osteoarticular allograft. Arch Orthop Trauma Surg 1996;115(2):100−3.

[19] Finkemeier CG. Bone-grafting and bone-graft substitutes. J Bone Joint Surg Am 2002; 84A(3):454−64.

[20] Kerr EJ III, et al. The use of osteo-conductive stem-cells allograft in lumbar interbody fusion procedures: an alternative to recombinant human bone morphogenetic protein. J Surg Orthop Adv 2011;20(3):193–7.

[21] Hollawell SM. Allograft cellular bone matrix as an alternative to autograft in hindfoot and ankle fusion procedures. J Foot Ankle Surg 2012;51(2):222–5.

[22] Lichte P, et al. Scaffolds for bone healing: concepts, materials and evidence. Injury—Int J Care Injured 2011;42(6):569–73.

[23] Seebach C, et al. Comparison of six bone-graft substitutes regarding to cell seeding efficiency, metabolism and growth behaviour of human mesenchymal stem cells (MSC) in vitro. Injury—Int J Care Injured 2010;41(7):731–8.

[24] Nguyen LH, et al. Vascularized bone tissue engineering: approaches for potential improvement. Tissue Eng Part BRev 2012;18(5):363–82.

[25] Place ES, et al. Synthetic polymer scaffolds for tissue engineering. Chem Soc Rev 2009; 38(4):1139–51.

[26] Nauth A, et al. Growth factors and bone regeneration: how much bone can we expect? Injury—Int J Care Injured 2011;42(6):574–9.

[27] Lissenberg-Thunnissen SN, et al. Use and efficacy of bone morphogenetic proteins in fracture healing. Int Orthop 2011;35(9):1271–80.

[28] Garrison KR, et al. Bone morphogenetic protein (BMP) for fracture healing in adults. Cochrane Database Syst Rev 2010;(6):1–155.

[29] Carragee EJ, Hurwitz EL, Weiner BK. A critical review of recombinant human bone morphogenetic protein-2 trials in spinal surgery: emerging safety concerns and lessons learned. Spine J 2011;11(6):471–91.

[30] Kuznetsov SA, et al. Circulating skeletal stem cells. J Cell Biol 2001;153(5):1133–9.

[31] Bianco P, et al. Postnatal skeletal stem cells. In: Klimanskaya I, Lanza R, editors. Adult stem cells. San Diego, CA: Elsevier Academic Press Inc; 2006. p. 117–48.

[32] Modder UI, Khosla S. Skeletal stem/osteoprogenitor cells: current concepts, alternate hypotheses, and relationship to the bone remodeling compartment. J Cell Biochem 2008; 103(2):393–400.

[33] Yang XBB, et al. Human osteoprogenitor bone formation using encapsulated bone morphogenetic protein 2 in porous polymer scaffolds. Tissue Eng 2004;10(7–8):1037–45.

[34] Sacchetti B, et al. Self-renewing osteoprogenitors in bone marrow sinusoids can organize a hematopoietic microenvironment. Cell 2007;131(2):324–36.

[35] Prockop DJ. Marrow stromal cells as steam cells for nonhematopoietic tissues. Science 1997;276(5309):71–4.

[36] Banfi A, et al. Proliferation kinetics and differentiation potential of ex vivo expanded human bone marrow stromal cells: implications for their use in cell therapy. Exp Hematol 2000; 28(6):707–15.

[37] Quarto R, et al. Repair of large bone defects with the use of autologous bone marrow stromal cells. New Engl J Med 2001;344(5):385–6.

[38] Bellantuono I, Aldahmash A, Kassem M. Aging of marrow stromal (skeletal) stem cells and their contribution to age-related bone loss. Biochimica Et Biophysica Acta—Molecular Basis Disease 2009;1792(4):364–70.

[39] Muraglia A, Cancedda R, Quarto R. Clonal mesenchymal progenitors from human bone marrow differentiate in vitro according to a hierarchical model. J Cell Sci 2000;113(7):1161–6.

[40] Jones EA, et al. Isolation and characterization of bone marrow multipotential mesenchymal progenitor cells. Arthritis Rheu 2002;46(12):3349–60.

[41] Pittenger MF, et al. Multilineage potential of adult human mesenchymal stem cells. Science 1999;284(5411):143−7.

[42] Caplan AI. Mesenchymal stem cells and gene therapy. Clin Orthop Relat Res 2000;(379): S67−70.

[43] Doulatov S, et al. Hematopoiesis: a human perspective. Cell Stem Cell 2012;10(2):120−36.

[44] Appelbaum FR. Hematopoietic-cell transplantation at 50. New Engl J Med 2007; 357(15):1472−5.

[45] Friedenstein AJ, et al. Heterotopic transplants of bone marrow — analysis of precursor cells for osteogenic and hematopoietic tissues. Transplantation 1968;6(2):230−47.

[46] Friedenstein AJ, Gorskaja UF, Kulagina NN. Fibroblast precursors in normal and irradiated mouse hematopoietic organs. Exp Hematol 1976;4(5):267−74.

[47] Friedenstein AJ. Precursor cells of mechanocytes. Int Rev Cytol Surv Cell Biol 1976;47:327−59.

[48] Caplan AI. Mesenchymal stem-cells. J Orthop Res 1991;9(5):641−50.

[49] Caplan AI. The mesengenic process. Clin Plast Surg 1994;21(3):429−35.

[50] Simmons PJ, Torokstorb B. Identification of stromal cell precursors in human bone-marrow by a novel monoclonal-antibody, stro-1. Blood 1991;78(1):55−62.

[51] Bianco P, Robey PG. Marrow stromal stem cells. J Clin Invest 2000;105(12):1663−8.

[52] Bartholomew A, et al. Mesenchymal stem cells suppress lymphocyte proliferation in vitro and prolong skin graft survival in vivo. Exp Hematol 2002;30(1):42−8.

[53] Le Blanc K, et al. Mesenchymal stem cells inhibit and stimulate mixed lymphocyte cultures and mitogenic responses independently of the major histocompatibility complex. Scand J Immunol 2003;57(1):11−20.

[54] Min AD, Theise ND. Prospects for cell-based therapies for liver disease. Panminerva Med 2004;46(1):43−8.

[55] Barry FP. Biology and clinical applications of mesenchymal stem cells. Birth Defects Res 2003;69(3):250−6.

[56] De Bari C, et al. Human synovial membrane-derived mesenchymal stem cells for skeletal muscle repair. Arthritis Rheu 2001;44(9):S101.

[57] Zuk PA, et al. Multilineage cells from human adipose tissue: implications for cell-based therapies. Tissue Eng 2001;7(2):211−28.

[58] Romanov YA, Svintsitskaya VA, Smirnov VN. Searching for alternative sources of postnatal human mesenchymal stem cells: candidate MSC-like cells from umbilical cord. Stem Cells 2003;21(1):105−10.

[59] Sarugaser R, et al. Human umbilical cord perivascular (HUCPV) cells: a source of mesenchymal progenitors. Stem Cells 2005;23(2):220−9.

[60] Campagnoli C, et al. Identification of mesenchymal stem/progenitor cells in human first-trimester fetal blood, liver, and bone marrow. Blood 2001;98(8):2396−402.

[61] Zhang Y, et al. Human placenta-derived mesenchymal progenitor cells support culture expansion of long-term culture-initiating cells from cord blood CD34(+) cells. Exp Hematol 2004;32(7):657−64.

[62] In't Anker PS, et al. Isolation of mesenchymal stem cells of fetal or maternal origin from human placenta. Stem Cells 2004;22(7):1338−45.

[63] Tsai MS, et al. Isolation of human multipotent mesenchymal stem cells from second-trimester amniotic fluid using a novel two-stage culture protocol. Human Reprod 2004;19(6):1450−6.

[64] Guillot PV, et al. Human first-trimester fetal MSC express pluripotency markers and grow faster and have longer telomeres than adult MSC. Stem Cells 2007;25(3):646—54.

[65] DiGirolamo CM, et al. Propagation and senescence of human marrow stromal cells in culture: a simple colony-forming assay identifies samples with the greatest potential to propagate and differentiate. Brit J Haematol 1999;107(2):275—81.

[66] Banfi A, et al. Replicative aging and gene expression in long-term cultures of human bone marrow stromal cells. Tissue Eng 2002;8(6):901—10.

[67] Miura M, et al. Accumulated chromosomal instability in murine bone marrow mesenchymal stem cells leads to malignant transformation. Stem Cells 2006;24(4):1095—103.

[68] Tarte K, et al. Clinical-grade production of human mesenchymal stromal cells: occurrence of aneuploidy without transformation. Blood 2010;115(8):1549—53.

[69] Prockop DJ, et al. Defining the risks of mesenchymal stromal cell therapy. Cytotherapy 2010;12(5):576—8.

[70] Dominici M, et al. Minimal criteria for defining multipotent mesenchymal stromal cells. The international society for cellular therapy position statement. Cytotherapy 2006; 8(4):315—7.

[71] Friedens Aj, et al. Precursors for fibroblasts in different populations of hematopoietic cells as detected by invitro colony assay method. Exp Hematol 1974;2(2):83—92.

[72] Castro-Malaspina H, et al. Characterisation of human bone marrow fibroblast colony-forming cells (CFU-F) and their progeny. Blood 1980;56(2):289—301.

[73] Jones EA, et al. Enumeration and phenotypic characterization of synovial fluid multipotential mesenchymal progenitor cells in inflammatory and degenerative arthritis. Arthritis Rheu 2004;50(3):817—27.

[74] De Bari C, et al. Mesenchymal multipotency of adult human periosteal cells demonstrated by single-cell lineage analysis. Arthritis Rheu 2006;54(4):1209—2121.

[75] De Bari C, et al. Multipotent mesenchymal stem cells from adult human synovial membrane. Arthritis Rheu 2001;44(8):1928—42.

[76] Tuli R, et al. Characterization of multipotential mesenchymal progenitor cells derived from human trabecular bone. Stem Cells 2003;21(6):681—93.

[77] Wickham MQ, et al. Multipotent stromal cells derived from the infrapatellar fat pad of the knee. Clin Orthop Relat Res 2003;(412):196—212.

[78] Salingcarnboriboon R, et al. Establishment of tendon-derived cell lines exhibiting pluripotent mesenchymal stem cell-like property. Exp Cell Res 2003;287(2):289—300.

[79] Dowthwaite GP, et al. The surface of articular cartilage contains a progenitor cell population. J Cell Sci 2004;117(6):889—97.

[80] Jones E, et al. Synovial fluid mesenchymal stem cells in health and early osteoarthritis: detection and functional evaluation at the single-cell level. Arthritis Rheu 2008; 58(6):1731—40.

[81] Morito T, et al. Synovial fluid-derived mesenchymal stem cells increase after intra-articular ligament injury in humans. Rheumatology 2008;ken114.

[82] Caplan AI. Mesenchymal stem cells: cell-based reconstructive therapy in orthopedics. Tissue Eng 2005;11(7—8):1198—211.

[83] Tuan RS. Stemming cartilage degeneration: adult mesenchymal stem cells as a cell source for articular cartilage tissue engineering. Arthritis Rheu 2006;54(10):3075—8.

[84] Alves AGL, et al. Cell-based therapies for tendon and ligament injuries. Vet Clin N Am— Equine Pract 2011;27(2):315.

[85] De Bari C, et al. A biomarker-based mathematical model to predict bone-forming potency of human synovial and periosteal mesenchymal stem cells. Arthritis Rheum 2008; 58(1):240—50.

[86] Sakaguchi Y, et al. Comparison of human stem cells derived from various mesenchymal tissues—superiority of synovium as a cell source. Arthritis Rheu 2005;52(8):2521—9.

[87] Guillot PV, et al. Comparative osteogenic transcription profiling of various fetal and adult mesenchymal stem cell sources. Differentiation 2008;76(9):946—57.

[88] Jaiswal N, et al. Osteogenic differentiation of purified, culture-expanded human mesenchymal stem cells in vitro. J Cell Biochem 1997;64(2):295—312.

[89] Bonewald LF. The amazing osteocyte. J Bone Miner Res 2011;26(2):229—38.

[90] Friedenstein AJ, Chailakhyan RK, Gerasimov UV. Bone-marrow osteogenic stem-cells—invitro cultivation and transplantation in diffusion-chambers. Cell Tissue Kinet 1987; 20(3):263—72.

[91] Friedenstein AJ, et al. Marrow micro-environment transfer by heterotopic transplantation of freshly isolated and cultured-cells in porous sponges. Exp Hematol 1982;10(2):217—27.

[92] De Bari C, et al. Skeletal muscle repair by adult human mesenchymal stem cells from synovial membrane. J Cell Biol 2003;160(6):909—18.

[93] Horwitz EM, et al. Transplantability and therapeutic effects of bone marrow-derived mesenchymal cells in children with osteogenesis imperfecta. Nat Med 1999;5(3):309—13.

[94] Horwitz EM, et al. Clinical responses to bone marrow transplantation in children with severe osteogenesis imperfecta. Blood 2001;97(5):1227—31.

[95] Horwitz EM, et al. Isolated allogeneic bone marrow-derived mesenchymal cells engraft and stimulate growth in children with osteogenesis imperfecta: implications for cell therapy of bone. Proc Natl Acad Sci USA 2002;99(13):8932—7.

[96] Le Blanc K, et al. Fetal mesenchymal stem-cell engraftment in bone after in utero transplantation in a patient with severe osteogenesis imperfecta. Transplantation 2005; 79(11):1607—14.

[97] Dominici M, et al. Donor cell-derived osteopoiesis originates from a self-renewing stem cell with a limited regenerative contribution after transplantation. Blood 2008;111(8):4386—91.

[98] Otsuru S, et al. Transplanted bone marrow mononuclear cells and MSCs impart clinical benefit to children with osteogenesis imperfecta through different mechanisms. Blood 2012;120(9):1933—41.

[99] Rose F, Oreffo ROC. Bone tissue engineering: hope vs hype. Biochem Biophys Res Commun 2002;292(1):1—7.

[100] Warnke PH, et al. Growth and transplantation of a custom vascularised bone graft in a man. Lancet 2004;364(9436):766—70.

[101] Marcacci M, et al. Stem cells associated with macroporous bioceramics for long bone repair: 6-to 7-year outcome of a pilot clinical study. Tissue Eng 2007;13(5):947—55.

[102] Dalby MJ, et al. The control of human mesenchymal cell differentiation using nanoscale symmetry and disorder. Nat Mater 2007;6(12):997—1003.

[103] D'Ippolito G, et al. Marrow-isolated adult multilineage inducible (MIAMI) cells, a unique population of postnatal young and old human cells with extensive expansion and differentiation potential. J Cell Sci 2004;117(14):2971—81.

[104] Poloni A, et al. Selection of CD271 + cells and human AB serum allows a large expansion of mesenchymal stromal cells from human bone marrow. Cytotherapy 2009;11(2):153—62.

[105] Bieback K, et al. Human alternatives to fetal bovine serum for the expansion of mesenchymal stromal cells from bone marrow. Stem Cells 2009;27(9):2331—41.

[106] Nimura A, et al. Increased proliferation of human synovial mesenchymal stem cells with autologous human serum. Arthritis Rheu 2008;58(2):501–10.

[107] Pountos I, Georgouli T, Giannoudis PV. The effect of autologous serum obtained after fracture on the proliferation and osteogenic differentiation of mesenchymal stem cells. Cell Mol Biol 2008;54(1):33–9.

[108] Tonti GA, Mannello F. From bone marrow to therapeutic applications: different behaviour and genetic/epigenetic stability during mesenchymal stem cell expansion in autologous and foetal bovine sera? Int J Dev Biol 2008;52(8):1023–32.

[109] Gronthos S, Simmons PJ. The growth-factor requirements of stro-1-positive human bone-marrow stromal precursors under serum-deprived conditions in-vitro. Blood 1995; 85(4):929–40.

[110] Hirata J, et al. Effect of platelet-derived growth-factor and bone marrow-conditioned medium on the proliferation of human-bone marrow-derived fibroblastoid colony-forming cells. Acta Haematol 1985;74(4):189–94.

[111] Reyes M, et al. Purification and ex vivo expansion of postnatal human marrow meso-dermal progenitor cells (Retracted article. See vol. 113, pg. 2370, 2009). Blood 2001; 98(9):2615–25.

[112] Bianchi G, et al. Ex vivo enrichment of mesenchymal cell progenitors by fibroblast growth factor 2. Exp Cell Res 2003;287(1):98–105.

[113] Sun Y, et al. Rescuing replication and osteogenesis of aged mesenchymal stem cells by exposure to a young extracellular matrix. Faseb J 2011;25(5):1474–85.

[114] Chen X, et al. Bioreactor expansion of human adult bone marrow-derived mesenchymal stem cells. Stem Cells 2006;24(9):2052–9.

[115] Sakai S, et al. Rotating three-dimensional dynamic culture of adult human bone marrow-derived cells for tissue engineering of hyaline cartilage. J Orthop Res 2009;27(4):517–21.

[116] Doucet C, et al. Platelet lysates promote mesenchymal stem cell expansion: a safety substi-tute for animal serum in cell-based therapy applications. J Cell Physiol 2005;205(2):228–36.

[117] Zaky SH, et al. Platelet lysate favours in vitro expansion of human bone marrow stromal cells for bone and cartilage engineering. J Tissue Eng Regen Med 2008;2(8):472–81.

[118] Niemeyer P, et al. Comparison of mesenchymal stem cells from bone marrow and adipose tissue for bone regeneration in a critical size defect of the sheep tibia and the influence of platelet-rich plasma. Biomaterials 2010;31(13):3572–9.

[119] Dhillon RS, et al. Platelet-rich plasma therapy—future or trend? Arthritis Res Ther 2012;14(4):219 [Epub ahead of print]

[120] Josse C, et al. Systematic chromosomal aberrations found in murine bone marrow-derived mesenchymal stem cells. Stem Cells Dev 2010;19(8):1167–73.

[121] Ben-David U, Mayshar Y, Benvenisty N. Large-scale analysis reveals acquisition of lineage-specific chromosomal aberrations in human adult stem cells. Cell Stem Cell 2011; 9(2):97–102.

[122] Sensebe L, Bourin P, Tarte K. Good manufacturing practices production of mesenchymal stem/stromal cells. Human Gene Ther 2011;22(1):19–26.

[123] Sensebe L, et al. Limited acquisition of chromosomal aberrations in human adult mesen-chymal stromal cells. Cell Stem Cell 2012;10(1):9–10.

[124] Alhadlaq A, Mao JJ. Tissue-engineered osteochondral constructs in the shape of an articu-lar condyle. J Bone Joint Surg—Am Vol 2005;87A(5):936–44.

[125] Reynard LN, Loughlin J. Genetics and epigenetics of osteoarthritis. Maturitas 2012; 71(3):200–4.

[126] Nesic D, et al. Cartilage tissue engineering for degenerative joint disease. Adv Drug Deliver Rev 2006;58(2):300−22.

[127] McGonagle D, et al. The anatomical basis for a novel classification of osteoarthritis and allied disorders. J Anat 2010;216(3):279−91.

[128] Gilbert JE. Current treatment options for the restoration of articular cartilage. Am J Knee Surg 1998;11(1):42−6.

[129] Minas T, Nehrer S. Current concepts in the treatment of articular cartilage defects. Orthopedics 1997;20(6):525−38.

[130] Hurst JM, et al. Rehabilitation following microfracture for chondral injury in the knee. Clin Sports Med 2010;29(2):257−65.

[131] Brittberg M, et al. Treatment of deep cartilage defects in the knee with autologous chondrocyte transplantation. New Engl J Med 1994;331(14):889−95.

[132] Vanlauwe JJE, et al. Characterized chondrocyte implantation in the patellofemoral joint an up to 4-year follow-up of a prospective cohort of 38 patients. Am J Sports Med 2012; 40(8):1799−807.

[133] Marlovits S, et al. Clinical and radiological outcomes 5 years after matrix-induced autologous chondrocyte implantation in patients with symptomatic, traumatic chondral defects. Am J Sports Med 2012;40(10):2273−80.

[134] Filardo G, et al. Matrix-assisted autologous chondrocyte transplantation for cartilage regeneration in osteoarthritic knees: results and failures at midterm follow-up. Am J Sports Med 2013;41(1):95−100.

[135] Couto DS, Perez-Breva L, Cooney CL. Regenerative medicine: learning from past examples. Tissue Eng Part A 2012;18(21−22):2386−93.

[136] Clar C, et al. Clinical and cost-effectiveness of autologous chondrocyte implantation for cartilage defects in knee joints: systematic review and economic evaluation. Health Technol Assess 2005;9(47):I-82.

[137] Stoddart MJ, et al. Cells and biomaterials in cartilage tissue engineering. Regen Med 2009;4(1):81−98.

[138] Dell'Accio F, De Bari C, Luyten FP. Molecular markers predictive of the capacity of expanded human articular chondrocytes to form stable cartilage in vivo. Arthritis Rheu 2001;44(7):1608−19.

[139] Brittberg M. Cell carriers as the next generation of cell therapy for cartilage repair a review of the matrix-induced autologous chondrocyte implantation procedure. Am J Sports Med 2010;38(6):1259−71.

[140] Noeth U, Steinert AF, Tuan RS. Technology insight: adult mesenchymal stem cells for osteoarthritis therapy. Nat Clin Pract Rheumatol 2008;4(7):371−80.

[141] De Bari C, Dell'Accio F, Luyten FP. Failure of in vitro-differentiated mesenchymal stem cells from the synovial membrane to form ectopic stable cartilage in vivo. Arthritis Rheu 2004;50(1):142−50.

[142] Dell'Accio F, Bari CD, Luyten FP. Microenvironment and phenotypic stability specify tissue formation by human articular cartilage-derived cells in vivo. Exp Cell Res 2003; 287(1):16.

[143] Kafienah W, et al. Three-dimensional cartilage tissue engineering using adult stem cells from osteoarthritis patients. Arthritis Rheu 2007;56(1):177−87.

[144] Acharya C, et al. Enhanced chondrocyte proliferation and mesenchymal stromal cells chondrogenesis in coculture pellets mediate improved cartilage formation. J Cell Physiol 2012;227(1):88−97.

[145] Khan IM, et al. Fibroblast growth factor 2 and transforming growth factor beta 1 induce precocious maturation of articular cartilage. Arthritis Rheu 2011;63(11):3417–27.

[146] Mochizuki T, et al. Higher chondrogenic potential of fibrous synovium- and adipose synovium-derived cells compared with subcutaneous fat-derived cells—distinguishing properties of mesenchymal stem cells in humans. Arthritis Rheu 2006;54(3):843–53.

[147] English A, et al. A comparative assessment of cartilage and joint fat pad as a potential source of cells for autologous therapy development in knee osteoarthritis. Rheumatology 2007;46(11):1676–83.

[148] Jones EA, et al. Synovial fluid mesenchymal stem cells in health and early osteoarthritis: detection and functional evaluation at the single-cell level. Arthritis Rheum 2008; 58(6):1731–40.

[149] Hegewald AA, et al. Hyaluronic acid and autologous synovial fluid induce chondrogenic differentiation of equine mesenchymal stem cells: a preliminary study. Tissue Cell 2004; 36(6):431–8.

[150] Williams R, et al. Identification and clonal characterisation of a progenitor cell subpopulation in normal human articular cartilage. PLoS ONE 2010;5:10.

[151] Koshino T, et al. Regeneration of degenerated articular cartilage after high tibial valgus osteotomy for medial compartmental osteoarthritis of the knee. Knee 2003; 10(3):229–36.

[152] Lafeber FPJG, et al. Unloading joints to treat osteoarthritis, including joint distraction. Curr Opin Rheumatol 2006;18(5):519–25.

[153] Intema F, et al. Tissue structure modification in knee osteoarthritis by use of joint distraction: an open 1-year pilot study. Ann Rheum Dis 2011;70(8):1441–6.

[154] Bos PK, et al. Cellular origin of neocartilage formed at wound edges of articular cartilage in a tissue culture experiment. Osteoarthritis Cartilage 2008;16(2):204–11.

[155] Lee DH, et al. Synovial fluid CD34(−) CD44(+) CD90(+) mesenchymal stem cell levels are associated with the severity of primary knee osteoarthritis. Osteoarthritis Cartilage 2012;20(2):106–9.

[156] Sekiya I, et al. Human mesenchymal stem cells in synovial fluid increase in the knee with degenerated cartilage and osteoarthritis. J Orthop Res 2012;30(6):943–9.

[157] Sheng Zhang, et al. Autologous synovial fluid enhances migration of mesenchymal stem cells from synovium of osteoarthritis patients in tissue culture system. J Orthop Res 2008;26(10):1413–8.

[158] Endres M, et al. Synovial fluid recruits human mesenchymal progenitors from subchondral spongious bone marrow. J Orthop Res 2007;25(10):1299–307.

[159] Mendelson A, et al. Chondrogenesis by chemotactic homing of synovium, bone marrow, and adipose stem cells in vitro. FASEB J Off Publ Fed Am Soc Exp Biol 2011; 25(10):3496–504.

[160] Lee CH, et al. Regeneration of the articular surface of the rabbit synovial joint by cell homing: a proof of concept study. Lancet 2010;376(9739):440–8.

[161] Kurth TB, et al. Functional mesenchymal stem cell niches in adult mouse knee joint synovium in vivo. Arthritis Rheu 2011;63(5):1289–300.

[162] Chow P. The rationale for the use of animal models in biomedical research. In: Chow PKH, Nq RTH, Oqden BE, editors. Using animal models in biomedical research. A primer for the investigator. Singapore: World Scientific Publishing Company; 2008. p. P2–10.

[163] Horner EA, et al. Long bone defect models for tissue engineering applications: criteria for choice. Tissue Eng Part B-Rev 2010;16(2):263–71.

[164] Horner E, Kirkham J, Yang X. Animal models. In: Polak JM, Harding SE, editors. Advances in tissue engineering. London: Imperial College Press; 2008. p. 763−80.

[165] Hattori H, et al. Bone formation using human adipose tissue-derived stromal cells and a biodegradable scaffold. J Biomed Mater Res Part B—Appl Biomater 2006;76B(1):230−9.

[166] Dennis JE, et al. Osteogenesis in marrow-derived mesenchymal cell porous ceramic composites transplanted subcutaneously: effect of fibronectin and laminin on cell retention and rate of osteogenic expression. Cell Transplant 1992;1(1):23−32.

[167] Isogai N, et al. Formation of phalanges and small joints by tissue-engineering. J Bone Joint Surg—Am Vol 1999;81A(3):306−16.

[168] Eyckmans J, Luyten FP. Species specificity of ectopic bone formation using periosteum-derived mesenchymal progenitor cells. Tissue Eng 2006;12(8):2203−13.

[169] Yang KGA, et al. The effect of synovial fluid from injured knee joints on in vitro chondrogenesis. Tissue Eng 2006;12(10):2957−64.

[170] Yang XB, et al. Biomimetic collagen scaffolds for human bone cell growth and differentiation. Tissue Eng 2004;10(7−8):1148−59.

[171] Jones E, Yang XB. Mesenchymal stem cells and bone regeneration: current status. Injury—Int J Care Injured 2011;42(6):562−8.

[172] Ashton BA, et al. Formation of bone and cartilage by marrow stromal cells in diffusion-chambers invivo. Clin Orthop Relat Res 1980;(151):294−307.

[173] Gundle R, Joyner CJ, Triffitt JT. Human bone tissue formation in-diffusion chamber culture in-vivo by bone-derived cells and marrow stromal fibroblastic cells. Bone 1995; 16(6):597−601.

[174] Partridge K, et al. Adenoviral BMP-2 gene transfer in mesenchymal stem cells: in vitro and in vivo bone formation on biodegradable polymer scaffolds. Biochem Biophys Res Commun 2002;292(1):144−52.

[175] Hillam R, et al. Comparison of physiological strains in the human skull and tibia. Bone 1996;19:686 686

[176] Hong L, Mao JJ. Tissue-engineered rabbit cranial suture from autologous fibroblasts and BMP2. J Dent Res 2004;83(10):751−6.

[177] Peltola MJ, et al. In vivo model for frontal sinus and calvarial bone defect obliteration with bioactive glass S53P4 and hydroxyapatite. J Biomed Mater Res 2001;58(3):261−9.

[178] Chen TM, et al. Reconstruction of calvarial bone defects using an osteoconductive material and post-implantation hyperbaric oxygen treatment. Mater Sci Eng C 2004;24(6−8): 855−60.

[179] Cooper GM, et al. Testing the critical size in calvarial bone defects: revisiting the concept of a critical-size defect. Plast Reconstr Surg 2010;125(6):1685−92.

[180] Ma D, et al. Reconstruction of rabbit critical-size calvarial defects using autologous bone marrow stromal cell sheets. Ann Plast Surg 2010;65(2):259−65.

[181] Tolli H, et al. Bioglass as a carrier for reindeer bone protein extract in the healing of rat femur defect. J Mater Sci Mater Med 2010;21(5):1677−84.

[182] Claes L, et al. Moderate soft tissue trauma delays new bone formation only in the early phase of fracture healing. J Orthop Res 2006;24(6):1178−85.

[183] Kokubu T, et al. Development of an atrophic nonunion model and comparison to a closed healing fracture in rat femur. J Orthop Res 2003;21(3):503−10.

[184] Einhorn TA, et al. The healing of segmental bone defects induced by demineralized bone matrix. A radiographic and biomechanical study. J Bone Joint Surg Am 1984;66(2):274−9.

[185] Sciadini MF, Johnson KD. Evaluation of recombinant human bone morphogenetic protein-2 as a bone-graft substitute in a canine segmental defect model. J Orthop Res 2000;18(2):289–302.

[186] Cook SD, et al. Effect of recombinant human osteogenic protein-1 on healing of segmental defects in nonhuman-primates. J Bone Joint Surg—Am Vol 1995;77A(5):734–50.

[187] Cancedda R, Giannoni P, Mastrogiacomo M. A tissue engineering approach to bone repair in large animal models and in clinical practice. Biomaterials 2007;28(29):4240–50.

[188] Bruder SP, et al. The effect of implants loaded with autologous mesenchymal stem cells on the healing of canine segmental bone defects. J Bone Joint Surg Am 1998;80(7):985–96.

[189] Kon E, et al. Autologous bone marrow stromal cells loaded onto porous hydroxyapatite ceramic accelerate bone repair in critical-size defects of sheep long bones. J Biomed Mater Res 2000;49(3):328–37.

[190] Bensaid W, et al. De novo reconstruction of functional bone by tissue engineering in the metatarsal sheep model. Tissue Eng 2005;11(5–6):814–24.

[191] Viateau V, et al. Induction of a barrier membrane to facilitate reconstruction of massive segmental diaphyseal bone defects: an ovine model. Vet Surg 2006;35(5):445–52.

[192] Mastrogiacomo M, et al. Reconstruction of extensive long bone defects in sheep using resorbable bioceramics based on silicon stabilized tricalcium phosphate. Tissue Eng 2006;12(5):1261–73.

[193] Hahn JA, et al. Double-plating of ovine critical sized defects of the tibia: a low morbidity model enabling continuous in vivo monitoring of bone healing. BMC Musculoskel Dis 2011;12:214.

[194] Xu JZ, et al. Repair of large segmental bone defects using bone marrow stromal cells with demineralized bone matrix. Orthop Surg 2009;1(1):34–41.

[195] Kinsella Jr. CR, et al. BMP-2-mediated regeneration of large-scale cranial defects in the canine: an examination of different carriers. Plast Reconstr Surg 2011;127(5):1865–73.

[196] Nukavarapu SP, Dorcemus DL. Osteochondral tissue engineering: current strategies and challenges. Biotechnol Adv 2012.

[197] Tokuhara Y, et al. Repair of experimentally induced large osteochondral defects in rabbit knee with various concentrations of Escherichia coli-derived recombinant human bone morphogenetic protein-2. Int Orthop 2010;34(5):761–7.

[198] Sun Y, et al. The regenerative effect of platelet-rich plasma on healing in large osteochondral defects. Int Orthop 2010;34(4):589–97.

[199] Jin LH, et al. Implantation of bone marrow-derived buffy coat can supplement bone marrow stimulation for articular cartilage repair. Osteoarthritis Cartilage 2011;19(12):1440–8.

[200] Nishino T, et al. Effect of gradual weight-bearing on regenerated articular cartilage after joint distraction and motion in a rabbit model. J Orthop Res 2010;28(5):600–6.

[201] Stevenson S. The immune response to osteochondral allografts in dogs. J Bone Joint Surg Am 1987;69(4):573–82.

[202] Zhu S, et al. NEL-like molecule-1-modified bone marrow mesenchymal stem cells/poly lactic-co-glycolic acid composite improves repair of large osteochondral defects in mandibular condyle. Osteoarthritis Cartilage 2011;19(6):743–50.

[203] Simon TM, Aberman HM. Cartilage regeneration and repair testing in a surrogate large animal model. Tissue Eng Part B Rev 2010;16(1):65–79.

[204] Giordano M, et al. Pridie's marrow stimulation technique combined with collagen matrix for cartilage repair. A study in a still growing sheep model. Int J Immunopathol Pharmacol 2011;24(1 Suppl 2):101–6.

[205] Zscharnack M, et al. Repair of chronic osteochondral defects using predifferentiated mesenchymal stem cells in an ovine model. Am J Sports Med 2010;38(9):1857–69.

[206] Ho ST, et al. The evaluation of a biphasic osteochondral implant coupled with an electrospun membrane in a large animal model. Tissue Eng Part A 2010;16(4):1123–41.

[207] Im GI, et al. A hyaluronate-atelocollagen/beta-tricalcium phosphate–hydroxyapatite biphasic scaffold for the repair of osteochondral defects: a porcine study. Tissue Eng Part A 2010;16(4):1189–200.

[208] Hunziker EB. Biologic repair of articular cartilage: defect models in experimental animals and matrix requirements. Clin Orthop Relat Res 1999;367(Suppl):S135–46.

[209] Hunziker EB, Quinn TM, Hauselmann HJ. Quantitative structural organization of normal adult human articular cartilage. Osteoarthritis Cartilage 2002;10(7):564.

[210] Shimomura K, et al. The influence of skeletal maturity on allogenic synovial mesenchymal stem cell-based repair of cartilage in a large animal model. Biomaterials 2010;31(31): 8004–11.

[211] Gelse K, et al. Cell-based resurfacing of large cartilage defects: long-term evaluation of grafts from autologous transgene-activated periosteal cells in a porcine model of osteoarthritis. Arthritis Rheum 2008;58(2):475–88.

[212] Gronthos S, et al. Molecular and cellular characterisation of highly purified stromal stem cells derived from human bone marrow. J Cell Sci 2003;116(9):1827–35.

[213] Tormin A, et al. CD146 expression on primary non-hematopoietic bone marrow stem cells correlates to in situ localization. Blood 2011;117(19):5067–77.

[214] Battula VL, et al. Isolation of functionally distinct mesenchymal stem cell subsets using antibodies against CD56, CD271, and mesenchymal stem cell antigen-1. Haematol Hematol J 2009;94(2):173–84.

[215] Mendez-Ferrer S, et al. Mesenchymal and haematopoietic stem cells form a unique bone marrow niche. Nature 2010;466(7308):829–34.

[216] Phinney DG, et al. Plastic adherent stromal cells from the bone marrow of commonly used strains of inbred mice: variations in yield, growth, and differentiation. J Cell Biochem 1999;72(4):570–85.

[217] Meirelles LdS, Nardi NB. Murine marrow-derived mesenchymal stem cell: isolation, in vitro expansion, and characterization. Br J Haematol 2003;123(4):702–11.

[218] Mendez-Ferrer S, et al. Haematopoietic stem cell release is regulated by circadian oscillations. Nature 2008;452(7186):442 U4

[219] Jones EA, et al. Optimization of a flow cytometry-based protocol for detection and phenotypic characterization of multipotent mesenchymal stromal cells from human bone marrow. Cytometry Part B Clin Cytometry 2006;70(6):391–9.

[220] Jones E, McGonagle D. Human bone marrow mesenchymal stem cells in vivo. Rheumatology 2008;47(2):126–31.

[221] Churchman SM, et al. Transcriptional profile of native CD271+ multipotential stromal cells: evidence for multiple fates, with prominent osteogenic and Wnt pathway signaling activity. Arthritis Rheu 2012;64(8):2632–43.

[222] Di Maggio N, et al. Fibroblast growth factor-2 maintains a niche-dependent population of self-renewing highly potent non-adherent mesenchymal progenitors through FGFR2c. Stem Cells 2012;30(7):1455–64.

[223] Baksh D, Davies JE, Zandstra PW. Adult human bone marrow-derived mesenchymal progenitor cells are capable of adhesion-independent survival and expansion. Exp Hematol 2003;31(8):723–32.

[224] Cuthbert R, et al. Single-platform quality control assay to quantify multipotential stromal cells in bone marrow aspirates prior to bulk manufacture or direct therapeutic use. Cytotherapy 2012;14(4):431−40.

[225] Qian H, Le Blanc K, Sigvardsson M. Primary mesenchymal stem and progenitor cells from bone marrow lack expression of CD44. J Biol Chem 2012.

[226] Haniffa MA, et al. Mesenchymal stem cells: the fibroblasts' new clothes? Haematologica 2009;94(2):258−63.

[227] Ishii M, et al. Molecular markers distinguish bone marrow mesenchymal stem cells from fibroblasts. Biochem Biophys Res Commun 2005;332(1):297−303.

[228] Halfon S, et al. Markers distinguishing mesenchymal stem cells from fibroblasts are down-regulated with passaging. Stem Cell Dev 2011;20(1):53−66.

[229] Sudo K, et al. Mesenchymal progenitors able to differentiate into osteogenic, chondrogenic, and/or adipogenic cells in vitro are present in most primary fibroblast-like cell populations. Stem Cells 2007;25(7):1610−7.

[230] Lorenz K, et al. Multilineage differentiation potential of human dermal skin-derived fibroblasts. Exp Dermatol 2008;17(11):925−32.

[231] Vaculik C, et al. Human dermis harbors distinct mesenchymal stromal cell subsets. J Invest Dermatol 2012;132(3):563−74.

[232] Haniffa MA, et al. Adult human fibroblasts are potent immunoregulatory cells and functionally equivalent to mesenchymal stem cells. J Immunol 2007;179(3):1595−604.

[233] Smith JR, Hayflick L. Variation in life-span of clones derived from human diploid cell strains. J Cell Biol 1974;62(1):48−53.

[234] Parsonage G, et al. Global gene expression profiles in fibroblasts from synovial, skin and lymphoid tissue reveals distinct cytokine and chemokine expression patterns. Thromb Haemostasis 2003;90(4):688−97.

[235] Akintoye SO, et al. Skeletal site-specific characterization of orofacial and iliac crest human bone marrow stromal cells in same individuals. Bone 2006;38(6):758−68.

[236] Maumus M, et al. Native human adipose stromal cells: localization, morphology and phenotype. Int J Obesity 2011;35(9):1141−53.

[237] Battula VL, et al. Prospective isolation and characterization of mesenchymal stem cells from human placenta using a frizzled-9-specific monoclonal antibody. Differ Res Biol Divers 2008;76(4):326−36.

[238] Hermida-Gomez T, et al. Quantification of cells expressing mesenchymal stem cell markers in healthy and osteoarthritic synovial membranes. J Rheumatol 2011;38(2):339−49.

[239] Noth U, et al. Multilineage mesenchymal differentiation potential of human trabecular bone-derived cells. J Orthop Res 2002;20(5):1060−9.

[240] Sakaguchi Y, et al. Suspended cells from trabecular bone by collagenase digestion become virtually identical to mesenchymal stem cells obtained from marrow aspirates. Blood 2004;104(9):2728−35.

[241] Jones E, et al. Large-scale extraction and characterization of CD271+ multipotential stromal cells from trabecular bone in health and osteoarthritis: implications for bone regeneration strategies based on uncultured or minimally cultured multipotential stromal cells. Arthritis Rheum 2010;62(7):1944−54.

[242] Cox G, et al. High abundance of CD271(+) multipotential stromal cells (MSCs) in intramedullary cavities of long bones. Bone 2012;50(2):510−7.

[243] Meirelles LD, Caplan AI, Nardi NB. In search of the in vivo identity of mesenchymal stem cells. Stem Cells 2008;26(9):2287−99.

[244] Crisan M, et al. A perivascular origin for mesenchymal stem cells in multiple human organs. Cell Stem Cell 2008;3(3):301–13.

[245] Bianco P, et al. "Mesenchymal" stem cells in human bone marrow (skeletal stem cells): a critical discussion of their nature, identity, and significance in incurable skeletal disease. Human Gene Ther 2010;21(9):1057–66.

[246] Dore-Duffy P, Cleary K. Morphology and properties of pericytes. In: Nag S, editor. Blood-brain and other neural barriers: reviews and protocols. Methods in molecular biology 2011;686: 49–68. doi: 10.1007/978-1-60761-938-3_2.

[247] Doherty MJ, et al. Vascular pericytes express osteogenic potential in vitro and in vivo. J Bone Miner Res 1998;13(5):828–38.

[248] Bianco P, et al. Bone marrow stromal stem cells: nature, biology, and potential applications. Stem Cells 2001;19(3):180–92.

[249] Short B, et al. Mesenchymal stem cells. Arch Med Res 2003;34(6):565–71.

[250] Shi S, Gronthos S. Perivascular niche of postnatal mesenchymal stem cells in human bone marrow and dental pulp. J Bone Miner Res 2003;18(4):696–704.

[251] Solovey AN, et al. Identification and functional assessment of endothelial P1H12. J Lab Clin Med 2001;138(5):322–31.

[252] Bardin N, et al. Identification of CD146 as a component of the endothelial junction involved in the control of cell–cell cohesion. Blood 2001;98(13):3677–84.

[253] Caplan AI. All MSCs are pericytes? Cell Stem Cell 2008;3(3):229–30.

[254] Meirelles LD, et al. MSC frequency correlates with blood vessel density in equine adipose tissue. Tissue Eng Part A 2009;15(2):221–9.

[255] Nagase T, et al. Analysis of the chondrogenic potential of human synovial stem cells according to harvest site and culture parameters in knees with medial compartment osteoarthritis. Arthritis Rheu 2008;58(5):1389–98.

[256] Caplan AI, Correa D. PDGF in bone formation and regeneration: new insights into a novel mechanism involving MSCs. J Orthop Res 2011;29(12):1795–803.

[257] Feng JF, et al. Dual origin of mesenchymal stem cells contributing to organ growth and repair. Proc Nat Acad Sci USA 2011;108(16):6503–8.

[258] Caplan AI, Correa D. The MSC: an injury drugstore. Cell Stem Cell 2011;9(1):11–5.

[259] Waterman RS, et al. A new mesenchymal stem cell (MSC) paradigm: polarization into a pro-inflammatory MSC1 or an immunosuppressive MSC2 phenotype. PLoS ONE 2010;5:4.

[260] Taylor SM, Jones PA. Multiple new phenotypes induced in 10t1/2-cells and 3t3-cells treated with 5-azacytidine. Cell 1979;17(4):771–9.

[261] Fukuda K. Development of regenerative cardiomyocytes from mesenchymal stem cells for cardiovascular tissue engineering. Artif Organs 2001;25(3):187–93.

[262] Hare JM, et al. A randomized, double-blind, placebo-controlled, dose-escalation study of intravenous adult human mesenchymal stem cells (prochymal) after acute myocardial infarction. J Am Coll Cardiol 2009;54(24):2277–86.

[263] Pittenger MF, Martin BJ. Mesenchymal stem cells and their potential as cardiac therapeutics. Circ Res 2004;95(1):9–20.

[264] Baron F, et al. Cotransplantation of mesenchymal stem cells might prevent death from graft-versus-host disease (GVHD) without abrogating graft-versus-tumor effects after HLA-Mismatched allogeneic transplantation following nonmyeloablative conditioning. Biol Blood Marrow Transplant 2010;16(6):838–47.

62   References

[265] Tolar J, Villeneuve P, Keating A. Mesenchymal stromal cells for graft-versus-host disease. Hum Gene Ther 2011;22(3):257–62.

[266] Lin Y, Hogan WJ. Clinical application of mesenchymal stem cells in the treatment and prevention of graft-versus-host disease. Adv Hematol 2011;2011:427863.

[267] Abumaree M, et al. Immunosuppressive properties of mesenchymal stem cells. Stem Cell Rev Rep 2012;8(2):375–92.

[268] Djouad F, et al. Mesenchymal stem cells: innovative therapeutic tools for rheumatic diseases. Nat Rev Rheumatol 2009;5(7):392–9.

[269] Duffy MM, et al. Mesenchymal stem cell effects on T-cell effector pathways. Stem Cell Res Ther 2011;2.

[270] Rosner J, et al. The potential for cellular therapy combined with growth factors in spinal cord injury. Stem cells Int 2012;2012:826754.

[271] Uccelli A, et al. Neuroprotective features of mesenchymal stem cells. Best Pract Res Clin Haematol 2011;24(1):59–64.

[272] Mazzini L, et al. Transplantation of mesenchymal stem cells in ALS. Prog Brain Res 2012;201:333–59.

[273] Lanza C, et al. Neuroprotective mesenchymal stem cells are endowed with a potent antioxidant effect in vivo. J Neurochem 2009;110(5):1674–84.

[274] Dai L-J, et al. The therapeutic potential of bone marrow-derived mesenchymal stem cells on hepatic cirrhosis. Stem Cell Res 2009;2(1):16–25.

[275] Aquino JB, et al. Mesenchymal stem cells as therapeutic tools and gene carriers in liver fibrosis and hepatocellular carcinoma. Gene Ther 2010;17(6):692–708.

[276] Majumdar MK, et al. Human marrow-derived mesenchymal stem cells (MSCs) express hematopoietic cytokines and support long-term hematopoiesis when differentiated toward stromal and osteogenic lineages. J Hematoth Stem Cell Res 2000;9(6):841–8.

[277] Nagasawa T. The chemokine CXCL12 and regulation of HSC and B lymphocyte development in the bone marrow niche. Osteoimmunology 2007;602:69–75.

[278] Kitaori T, et al. Stromal cell-derived factor 1/cxcr4 signaling is critical for the recruitment of mesenchymal stem cells to the fracture site during skeletal repair in a mouse model. Arthritis Rheu 2009;60(3):813–23.

[279] Son BR, et al. Migration of bone marrow and cord blood mesenchymal stem cells in vitro is regulated by stromal-derived factor-1-CXCR4 and hepatocyte growth factor-c-met axes and involves matrix metalloproteinases. Stem Cells 2006;24(5):1254–64.

[280] Kollet O, et al. Regulatory cross talks of bone cells, hematopoietic stem cells and the nervous system maintain hematopoiesis. Inflamm Allergy Drug Targ 2012;11(3):170–80.

[281] Schaumann DHS, et al. VCAM-1-positive stromal cells from human bone marrow producing cytokines for B lineage progenitors and for plasma cells: SDF-1, flt3L, and BAFF. Mol Immunol 2007;44(7):1606–12.

[282] Pillai M, Torok-Storb B, Iwata M. Expression and function of IL-7 receptors in marrow stromal cells. Leukemia Lymphoma 2004;45(12):2403–8.

[283] Wynn RF, et al. A small proportion of mesenchymal stem cells strongly expresses functionally active CXCR4 receptor capable of promoting migration to bone marrow. Blood 2004;104(9):2643–5.

[284] Hung SC, et al. Isolation and characterization of size-sieved stem cells from human bone marrow. Stem Cells 2002;20(3):249–58.

[285] Zhong W, et al. In vivo comparison of the bone regeneration capability of human bone marrow concentrates vs. platelet-rich plasma. PLoS ONE 2012;7:7.

[286] Gessmann J, et al. Regenerate augmentation with bone marrow concentrate after traumatic bone loss. Orthop Rev 2012;4(1):e14.

[287] Iafrati MD, et al. Early results and lessons learned from a multicenter, randomized, double-blind trial of bone marrow aspirate concentrate in critical limb ischemia. J Vasc Surg 2011;54(6):1650−8.

[288] Gigante A, et al. Use of collagen scaffold and autologous bone marrow concentrate as a one-step cartilage repair in the knee: histological results of second-look biopsies at 1 year follow-up. Int J Immunopath Pharmacol 2011;24(1 Suppl 2):69−72.

[289] Hendrich C, et al. Safety of autologous bone marrow aspiration concentrate transplantation: initial experiences in 101 patients. Orthop Rev 2009;1(2):e32.

[290] Kasten P, et al. Instant stem cell therapy: characterization and concentration of human mesenchymal stem cells in vitro. Eur Cells Mater 2008;16:47−55.

[291] Hernigou P, et al. Percutaneous autologous bone-marrow grafting for nonunions—influence of the number and concentration of progenitor cells. J Bone Joint Surg—Am Vol 2005;87A(7):1430−7.

[292] Hernigou P, et al. The use of percutaneous autologous bone marrow transplantation in nonunion and avascular necrosis of bone. J Bone Joint Surg—Br Vol 2005;87B(7):896−902.

[293] Park I-H, Micic ID, Jeon I-H. A study of 23 unicameral bone cysts of the calcaneus: open chip allogeneic bone graft versus percutaneous injection of bone powder with autogenous bone marrow. Foot Ankle Int 2008;29(2):164−70.

[294] Lin K, et al. Characterization of adipose tissue-derived cells isolated with the celution (TM) system. Cytotherapy 2008;10(4):417−26.

[295] Hicok KC, Hedrick MH. Automated isolation and processing of adipose-derived stem and regenerative cells. In: Gimble JM, Bunnell BA, editors. Adipose-derived stem cells: methods and protocols. Methods in molecular biology 2011; 702:87−105. doi: 10.1007/978-1-61737-960-4_8.

[296] Nauth A, et al. Growth factors: beyond bone morphogenetic proteins. J Orthop Trauma 2010;24(9):543−6.

[297] Kawaguchi H, et al. A local application of recombinant human fibroblast growth factor 2 for tibial shaft fractures: a randomized, placebo-controlled trial. J Bone Miner Res 2010; 25(12):2459−67.

[298] Myers TJ, et al. Systemically delivered insulin-like growth factor-I enhances mesenchymal stem cell-dependent fracture healing. Growth Factors 2012;30(4):230−41.

[299] Granero-Molto F, et al. Regenerative effects of transplanted mesenchymal stem cells in fracture healing. Stem Cells 2009;27(8):1887−98.

[300] Murata K, et al. Stromal cell-derived factor 1 regulates the actin organization of chondrocytes and chondrocyte hypertrophy. PLoS ONE 2012;7:5.

[301] Zhang W, et al. The use of type 1 collagen scaffold containing stromal cell-derived factor-1 to create a matrix environment conducive to partial-thickness cartilage defects repair. Biomaterials 2013;34(3):713−23.

[302] Warnke PH, et al. Man as living bioreactor: fate of an exogenously prepared customized tissue-engineered mandible. Biomaterials 2006;27(17):3163−7.

[303] Torroni A. Engineered bone grafts and bone flaps for maxillofacial defects: state of the art. J Oral Maxillofacial Surg 2009;67(5):1121−7.

[304] Masquelet AC, Begue T. The concept of induced membrane for reconstruction of long bone defects. Orthop Clin N Am 2010;41(1):27.

[305] Giannoudis PV, et al. Masquelet technique for the treatment of bone defects: tips-tricks and future directions. Injury—Int J Care Injured 2011;42(6):591–8.

[306] Pelissier P, et al. Induced membranes secrete growth factors including vascular and osteoinductive factors and could stimulate bone regeneration. J Orthop Res 2004;22(1):73–9.

[307] Guda T, et al. Guidedbone regeneration in long-bone defects with a structural hydroxyapatite graft and collagen membrane. Tissue Eng Part A 2012 [in press, Epub ahead of print].

[308] El Backly RM, et al. A platelet-rich plasma-based membrane as a periosteal substitute with enhanced osteogenic and angiogenic properties: a new concept for bone repair. Tissue Eng Part A 2013;19(1–2):152–65.

[309] Bensidhoum M, et al. Homing of in vitro expanded Stro-1(−) or Stro-1(+) human mesenchymal stem cells into the NOD/SCID mouse and their role in supporting human CD34 cell engraftment. Blood 2004;103(9):3313–9.

[310] Zhang YM, Adachi Y, et al. Simultaneous injection of bone marrow cells and stromal cells into bone marrow accelerates hematopoiesis in vivo. Stem cells 2004;22(7):1256–62.

[311] Kushida T, Ueda Y, et al. Allogeneic intra-bone marrow transplantation prevents rheumatoid arthritis in SKG/Jcl mice. J Autoimmun 2009;32(3–4):216–22.

[312] Deal C. Future therapeutic targets in osteoporosis. Curr Opin Rheumatol 2009;21(4): 380–5.

[313] Benisch P, et al. The transcriptional profile of mesenchymal stem cell populations in primary osteoporosis is distinct and shows overexpression of osteogenic inhibitors. PLoS ONE 2012;7:9.

[314] Ke HZ, et al. Sclerostin and dickkopf-1 as therapeutic targets in bone diseases. Endocrine Rev 2012;33(5):747–83.

[315] Gambardella A, et al. Glycogen synthase kinase-3 alpha/beta inhibition promotes in vivo amplification of endogenous mesenchymal progenitors with osteogenic and adipogenic potential and their differentiation to the osteogenic lineage. J Bone Miner Res 2011;26(4): 811–21.

[316] Luo J. Glycogen synthase kinase 3 beta (GSK3 beta) in tumorigenesis and cancer chemotherapy. Cancer Lett 2009;273(2):194–200.

[317] Klamer G, et al. Using small molecule GSK3 beta inhibitors to treat inflammation. Current Med Chem 2010;17(26):2873–81.

[318] Hare JM, et al. Comparison of allogeneic vs autologous bone marrow-derived mesenchymal stem cells delivered by transendocardial injection in patients with ischemic cardiomyopathy the poseidon randomized trial. J Am Med Assoc 2012;308(22):2369–79.

[319] Traverse JH, et al. Effect of the use and timing of bone marrow mononuclear cell delivery on left ventricular function after acute myocardial infarction the time randomized trial. J Am Med Assoc 2012;308(22):2380–9.

[320] MacMillan ML, et al. Transplantation of ex-vivo culture-expanded parental haploidentical mesenchymal stem cells to promote engraftment in pediatric recipients of unrelated donor umbilical cord blood: results of a phase I–II clinical trial. Bone Marrow Transplant 2009;43(6):447–54.

[321] Prasad VK, et al. Efficacy and safety of ex vivo cultured adult human mesenchymal stem cells (prochymal (TM)) in pediatric patients with severe refractory acute graft-versus-host disease in a compassionate use study. Biol Blood Marrow Transplant 2011;17(4): 534–41.

[322] Ra JC, et al. Safety of intravenous infusion of human adipose tissue-derived mesenchymal stem cells in animals and humans. Stem Cells Dev 2011;20(8):1297–308.

[323] Lu D, et al. Comparison of bone marrow mesenchymal stem cells with bone marrow-derived mononuclear cells for treatment of diabetic critical limb ischemia and foot ulcer: a double-blind, randomized, controlled trial. Diabetes Res Clin Pract 2011;92(1):26–36.

[324] Zhao D, et al. Treatment of early stage osteonecrosis of the femoral head with autologous implantation of bone marrow-derived and cultured mesenchymal stem cells. Bone 2012; 50(1):325–30.

[325] Wakitani S, et al. Safety of autologous bone marrow-derived mesenchymal stem cell transplantation for cartilage repair in 41 patients with 45 joints followed for up to 11 years and 5 months. J Tissue Eng Regen Med 2011;5(2):146–50.

[326] Karussis D, et al. Safety and immunological effects of mesenchymal stem cell transplantation in patients with multiple sclerosis and amyotrophic lateral sclerosis. Archives Neurol 2010;67(10):1187–94.

[327] Connick P, et al. Autologous mesenchymal stem cells for the treatment of secondary progressive multiple sclerosis: an open-label phase 2a proof-of-concept study. Lancet Neurol 2012;11(2):150–6.

[328] Chase LG, et al. A novel serum-free medium for the expansion of human mesenchymal stem cells. Stem Cell Res Ther 2010;1(1):8.

[329] Chase LG, et al. Development and characterization of a clinically compliant xeno-free culture medium in good manufacturing practice for human multipotent mesenchymal stem cells. Stem cells Trans Med 2012;1(10):750–8.

[330] Majumdar MK, et al. Isolation, characterization, and chondrogenic potential of human bone marrow-derived multipotential stromal cells. J Cell Physiol 2000;185(1):98–106.

[331] Quirici N, et al. Isolation of bone marrow mesenchymal stem cells by anti-nerve growth factor receptor antibodies. Exp Hematol 2002;30(7):783–91.

[332] Deschaseaux F, et al. Direct selection of human bone marrow mesenchymal stem cells using an anti-CD49a antibody reveals their CD45(med,low) phenotype. Br J Haematol 2003;122(3):506–17.

[333] Lee RH, et al. The CD34-like protein PODXL and alpha 6-integrin (CD49f) identify early progenitor MSCs with increased clonogenicity and migration to infarcted heart in mice. Blood 2009;113(4):816–26.

[334] Boiret N, et al. Characterization of nonexpanded mesenchymal progenitor cells from normal adult human bone marrow. Exp Hematol 2005;33(2):219–25.

[335] Aslan H, et al. Osteogenic differentiation of noncultured immunoisolated bone marrow-derived CD105+ cells. Stem Cells 2006;24(7):1728–37.

[336] Gronthos S, et al. A novel monoclonal antibody (STRO-3) identifies an isoform of tissue nonspecific alkaline phosphatase expressed by multipotent bone marrow stromal stem cells. Stem Cells Dev 2007;16(6):953–63.

[337] Martinez C, et al. Human bone marrow mesenchymal stromal cells express the neural ganglioside GD2: a novel surface marker for the identification of MSCs. Blood 2007;109(10): 4245–8.

[338] Gang EJ, et al. SSEA-4 identifies mesenchymal stem cells from bone marrow. Blood 2007;109(4):1743–51.

[339] Veyrat-Masson R, et al. Mesenchymal content of fresh bone marrow: a proposed quality control method for cell therapy. Br J Haematol 2007;139(2):312–20.

[340] Bühring H-J, et al. Novel markers for the prospective isolation of human MSC. In Kanz L, Weisel KC, Dick JE, et al. editors. Hematopoietic stem cells VI. 6th Biennial International Symposium and Workshop on Hematopoietic Stem Cells. Germany: Univ Tubingen,

Tubingen. Annals of the New York academy of sciences 2007;262–271. doi: 10.1196/annals.1392.000.

[341] Kastrinaki M-C, et al. Isolation of human bone marrow mesenchymal stem cells using different membrane markers: comparison of colony/cloning efficiency, differentiation potential, and molecular profile. Tissue Eng Part C—Methods 2008;14(4):333–9.

[342] Delorme B, et al. Specific plasma membrane protein phenotype of culture-amplified and native human bone marrow mesenchymal stem cells. Blood 2008;111(5):2631–5.

[343] Sorrentino A, et al. Isolation and characterization of CD146(+) multipotent mesenchymal stromal cells. Exp Hematol 2008;36(8):1035–46.

[344] Bae S, et al. Fibroblast activation protein alpha identifies mesenchymal stromal cells from human bone marrow. Br J Haematol 2008;142(5):827–30.

[345] Gronthos S, et al. Heat shock protein-90 beta is expressed at the surface of multipotential mesenchymal precursor cells: generation of a novel monoclonal antibody, STRO-4, with specificity for mesenchymal precursor cells from human and ovine tissues. Stem Cells Dev 2009;18(9):1253–61.

[346] Sobiesiak M, et al. The mesenchymal stem cell antigen MSCA-1 is identical to tissue non-specific alkaline phosphatase. Stem Cells Dev 2010;19(5):669–77.

[347] Kuci S, et al. CD271 antigen defines a subset of multipotent stromal cells with immunosuppressive and lymphohematopoietic engraftment-promoting properties. Haematologica 2010;95(4):651–9.

[348] Maijenburg MW, et al. The composition of the mesenchymal stromal cell compartment in human bone marrow changes during development and aging. Haematologica 2012; 97(2):179–83.

[349] Yang XB, Tare RS, et al. Induction of human osteoprogenitor chemotaxis, proliferation, differentiation, and bone formation by osteoblast stimulating factor-1/pleiotrophin: osteo-conductive biomimetic scaffolds for tissue engineering. J Bone Miner Res 2003;18(1): 47–57.